AN ECONOMIC ANALYSIS OF TILAPIA FARMING IN DEVELOPING COUNTRIES

Edited by:

Navghan Mahida

Delve
PUBLISHING

www.delvepublishing.com

An Economic Analysis of Tilapia Farming in Developing Countries

Navghan Mahida

Delve Publishing
224 Shoreacres Road
Burlington, ON L7L 2H2
Canada
www.delvepublishing.com
Email: orders@arclereducation.com

ISBN: 978-1-77469-427-5 (Hardcover)

ABOUT THE EDITOR

Dr. Navghan Mahida is a self-motivated fisheries scholar by qualification with over five years of work experience, including over 9 training programs and work exposure, possesses qualities of leadership, good communication & interpersonal skills, to do attitude, and ability to quickly adapt new environments. He is working as Assistant Professor in the Department of Animal Husbandry, Dairy and Fisheries, Parul University Vadodara-Gujarat. He had completed B.F.Sc (Bachelor of Fisheries Science) in 2012 from CoF-Veraval, Junagadh Agricultural University, Junagadh. He pursued M.F.Sc in 2014 and Ph.D. in 2019 from the Central Institute of Fisheries Education-Mumbai. He started his career as Fisheries Officer in the Government of Gujarat in 2013 and further worked in MNC Shoreline Aquaculture as Assistant Manager. He specialized in fisheries economics with commendable academic records and achievements. He has published 25 research papers in national and international peer-reviewed journals so far, including conduction and participation in conferences, conventions, symposiums, seminars, training, etc. He had written souvenirs for the students' convention programs and three popular articles. He is a lifetime member of fisheries and life sciences, the European Journal of fisheries. He is also a reputed editor in the journal of fisheries and life sciences, PLOS ONE San Francisco, California, USA, Asian Journal of Economics, Extension and Social Science Thailand etc.

TABLE OF CONTENTS

LIST OF FIGURES

PREFACE

Tilapia is one of the fish species that can be classified as a solution to the food crisis due to population growth. It is being grown in different places like reservoirs, ponds, etc. They can easily live in diverse climatic settings and the cost incurred in the management of tilapia is minimum. Due to its affordable cost, it is being consumed by all the sections of society with no restrictions. Most developing countries adopted tilapia in boosting their economy by either exporting it or consuming it. There are several benefits of tilapia and farmers are taking initiatives in growing it in inland aquacultures and marine settings. The primary economic benefits of tilapia can be observed in the rural communities mostly and this can be due to several factors. The tilapia culture can provide employment to the rural communities and they can easily invest in this business with limited hesitation. The food requirements of this species are minimal as they can consume natural foods that include fruit peel, rice bran, etc. This quality of tilapia helps the farmers in reducing the cost of production as there is no need to purchase commercial feed from the companies at higher costs. This species can be chosen for successful aquaculture and this can be attributed to its disease resistance. It can easily tolerate poor water quality that is common in the developing countries.

The yields in the tilapia culture can be dependent on diverse factors and this can be considered as a function of inputs that made into the manufacturing or production process. This species can be easily grown with other species and this forms a foundation for the polyculture of tilapia. Aquaponics can be employed in increasing productivity and reducing costs. Genetically improved tilapia farming or GIFT can be considered as an important technique that can fetch farmers several economic benefits. The selective breeding in GIFT can be employed to reap the economic fruits of tilapia aquaculture in developing and poor countries. Bangladesh, India, Sri Lanka and Thailand are some of the countries that get benefitted by tilapia aquaculture. The culture of tilapia in these countries is described in this book. Climate change has affected many businesses and yet tilapia aquaculture is beneficial as it can sustain the rough conditions that persist due to global warming.

This book can be used as a handbook and guide for postgraduates, graduates and researchers in the field of aquaculture that mainly focuses on tilapia.

INTRODUCTION

The demand for tilapia due to population needs can be due to its beneficial traits that can help all the sections of the society irrespective of their financial condition. It is next to carp in terms of production and the best if we consider the costs that can be incurred in tilapia culture. It is being called as an aquatic chicken as the poor people can easily afford this fish. This fish can be grown across diverse culture systems and hence farmers can select a profitable option. Tilapia provide a livelihood for several people and this can be in terms of income generation, employment and a source of animal protein with minimal cost. Tilapia species is believed to be originated from the African subcontinent and it has become a favorite and affordable food for many people across the globe. There are about 200 tilapia species that are being grown in various places and this shows their diversity. The technology that is being used in tilapia culture is shared with several Asian countries and specific attention from the developing countries made this species popular. This species can be grown easily on a large scale and small scale and this can be attributed to their adaptability to rough conditions. There are some limitations in tilapia production that hampers their growth. They lay limited eggs and asynchronous spawning and these factors can reduce the production rates. This species can easily breed in any culture system and this helps farmers to not purchase expensive hormones from the market. Fry can be produced in the process of tilapia culture and the farmers can easily sell these by-products with no additional investments. This species can offer us a good and fast turnover than other species and this makes tilapia culture beneficial in developing countries where economies are buoyant.

Tilapia culture helps farmers who are having limited resources, especially in the rural areas as this culture demands minimal investments. As this species survive on natural foods and even sustains worse water quality, it can be a better choice for rural entrepreneurs. The flavor and taste of tilapia species attract many consumers irrespective of financial disparities. The disease resistance property of tilapia can lower the financial burden among rural farmers. The skin of tilapia can be used for various purposes like sofas, shoes, etc. In this book, the economic benefits of tilapia are described. The dynamics of tilapia culture in developing countries is explained. The effect of climate change on tilapia production is discussed.

CHAPTER 1

INTRODUCTION TO TILAPIA

CONTENTS

1.1 INTRODUCTION

While tilapia has been cultivated since the early 1950s, it has recently grown in popularity and is now second only to carp in terms of global production (Figure 1.1). As a result, it's also known as aquatic chicken. Tilapia farming has now extended throughout a diverse spectrum of culture systems, from small ditches to big ponds and reservoirs, in both fresh and saltwater, and from peri-urban to rural settings(Boyd, n.d.). Tilapia is important for food security since it is consumed by the poor, particularly those who live in rural regions, as well as those who live in cities and buy their food from supermarkets. It now directly or indirectly supports these people's livelihoods by supplying low-cost animal protein, creating jobs, and earning revenue. Although tilapia originated in Africa, they are now found in most Asian and Latin American countries. Despite the fact that there are over 200 species of tilapia accessible for culture in different parts of the world, the Nile tilapia or *Oreochromis niloticus* became a focus species. Nile tilapia has a unique history in Thailand. It was first handed to the Royal Palace as a gift from the Japanese Emperor in 1965 (R. Bhujel, 2008). Many Thais see tilapia as a valuable fish, owing to the fact that it is supposed to be the King's fish. Once the fish were received, the ease of breeding and culture was rapidly realized, and research proceeded to create improved and low-cost culture procedures.

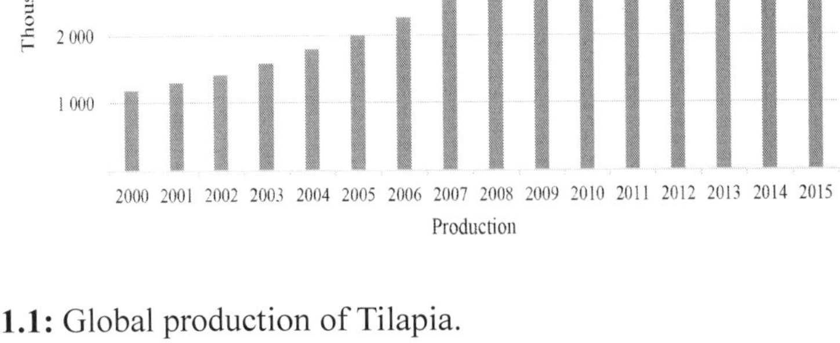

Figure 1.1: Global production of Tilapia.

Source: https://www.researchgate.net/profile/RohanaSubasinghe/publication/318997303/figure/fig4/AS:525462829006849@1502291509468/Global-production-of-tilapia-from-2000-2015_W640.jpg

Simple and low-cost techniques have a high percentage of adoption and success in poor countries. Because tilapia can survive in harsh environments, it has become a popular fish among the general public. Shrimp, catfish, and snakehead are all grown by wealthy farmers. When shrimp farming was decimated by disease, tilapia began to attract even wealthier growers as a viable alternative. Since then, tilapia has been seen as a potential species that may be maintained and produced in a variety of culture systems, from backyard ponds to intensively managed tanks and ponds. Tilapia can rely on plankton as a feed, which can be produced simply by fertilizing or manuring ponds with readily available chemical fertilizers used for rice, such as urea, triple superphosphate, etc., Farmers can also use rice bran, oilcakes, and other by-products from their crops as fish feed to boost output. As a result, tilapia has gained appeal not just in Thailand, but also in other countries in the region, such as China, Laos, the Philippines, Vietnam, Taiwan, Malaysia, and Indonesia, and people no longer regard it as an eccentric species.

1.2 CULTURE SYSTEMS

They breed naturally in culture systems; thus, farmers don't need to inject any hormones to breed them, unlike carp, catfish, and other species. Fry can also be produced as a by-product, reducing the need for farmers to purchase and transport fry from other sources and allowing them to earn money by selling fry in addition to table fish. Tilapia have a fast turnover because of their strong growth and survival rates. They are appropriate for resource-poor farmers in rural locations since they may be farmed with little input or expenditure. They eat a wide range of natural foods, including algae, detritus, and farm wastes like rice bran, oilcakes, vegetable leaves, and fruit peel. This implies that farmers rarely need to purchase commercial feed from outside sources. Because their muscles are white and lack intramuscular bones, they are referred to as boneless flesh. Their meat has a wonderful flavor and a pleasant taste that many people enjoy. They are disease-resistant fish, which means growers face less hazards. They may grow and breed in a variety of settings, including fresh or brackish waters in tiny ditches, rice fields, ponds, cages, tanks, or raceways, and they can endure poor water quality, such as low dissolved oxygen (DO), high ammonia, and high temperature. They can be used in monocultures as well as polycultures with a variety of other species, including carp, catfish, prawns, and others.

Tilapia became the second most significant species in terms of production volume only after carp in 2005, surpassing salmonid production. Annual production reached above 3.5 million t in 2011 and 3.7 million t in 2012(R. C. Bhujel, 2014). In 2013, it is estimated to be close to 4 million tons. With over 1 million tons, China remains the leading producer; nevertheless, other nations such as Egypt and Indonesia have experienced rapid growth(Aich, Paul, Choudhury, & Saha, 2022). Egypt has been substantially increasing production since the mid-1990s. The tilapia business in Thailand and the Philippines is continuously expanding. Tilapia feed at a low trophic level and have an eating behavior that falls midway between herbivore and omnivore. Phytoplankton and zooplankton, algae, some aquatic macrophytes, benthic invertebrates, other fish larvae, detritus, and decomposing organic waste are among the natural foods they ingest. Because they can easily capture plankton from the water, tilapia is often referred to as filter feeders. They eat live benthic invertebrates as well as bacteria-laden debris. Tilapia, on the other hand, do not disrupt the pond's bottom like common carp do. Tilapia are not piscivorous, however their fry and fingerlings may eat other fish larvae. Tilapia eat invertebrates in the water column as well. Even when tilapia is heavily supplemented, natural food may account for 30–50% of their growth(Groenewald, n.d.). Even when fish are carefully grown, especially in outdoor settings, the nutritional value of the natural food supply in ponds is crucial.

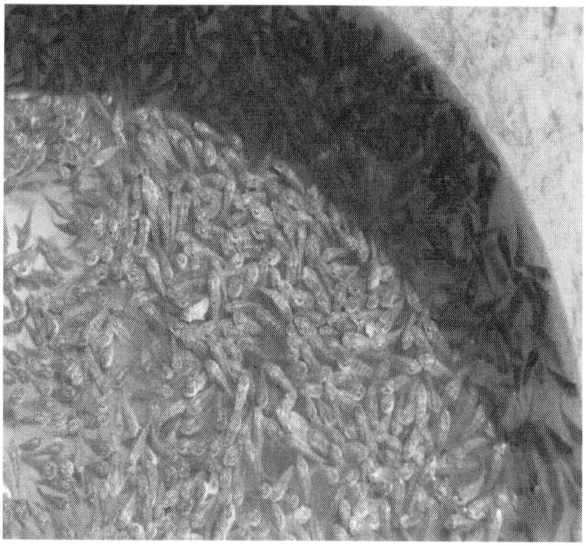

Figure 1.2: Tilapia breeding.

Source: https://www.worldfishcenter.org/sites/default/files/styles/large/public/2021-03/9120922867_70c6ceb034_k-1024x684.jpg?itok=gjI6FN70

1.3 BREEDING

Tilapia breed (Figure 1.2) easily in captivity and do not require hormone injections, which was once thought to be their greatest advantage over other species. Poor farmers could easily embrace tilapia farming and maintain their family-scale fish ponds by harvesting larger fish partially and leaving younger ones to thrive. Farmers later discovered that early breeding behavior restricts the growth of supplied fish. To overcome the difficulty of breeding in food fish production systems, many strategies were attempted, and some of these approaches are still used in various regions of the world. The culture of all male or primarily male fry has proven to be the most effective and commonly used of these cultural strategies. Farmers want to raise solely male tilapia to avoid breeding, as well as because males grow quicker than females. When fish are kept in cages, the eggs fall through the mesh net to the bottom, preventing the females from collecting and incubating them. Similarly, some farmers raise tilapia with carnivorous fish like catfish, sea bass, largemouth bass, and others so that they can consume the tilapia eggs and larvae before they grow any larger. These strategies each have their own set of advantages and disadvantages. Females still lose energy essential for reproduction when mixed-sex fry are cultured in cages and the egg falls through the cage. If half of the fry are female, a lot of energy is wasted. Male tilapia can grow twice as quickly as female tilapia. When half of the fish are smaller, total productivity is lower than when an all-male population is grown.

Cultivating tilapia with carnivorous fish will result in a comparable loss, however, compensation can be achieved through the sale of carnivorous fish in this situation. Farmers must be cautious about the density and size of carnivorous species stocked, as they may exert stress on stocked tilapia and potentially cause injury, resulting in stress that may favor disease outbreaks. Nonetheless, it was rapidly recognized that there was a problem and that the tilapia culture was not growing as swiftly as predicted. This prompted researchers to launch a research program targeted at improving the mass fry production techniques based on hormonal sex reversal, which had already been adopted by thousands of hatchery operators across Asia and beyond. Although there are four additional ways for producing primarily males in a fry population, hormonal sex reversal is the most frequently used and has spread to practically every country where tilapia is grown. Other approaches are still used with varying degrees of effectiveness, and each has its own set of advantages and disadvantages. Manual sexing by examining the genital papilla, for example, necessitates expert labor and is time-consuming. Hybridization between two species that generate mostly male offspring, such as Nile tilapia or Mozambique females crossed with blue or Zanzibar males, is another option. However, the results are inconsistent, and obtaining an all-male fry population is difficult, with only about 80% of male fry populations being attained on average(Tseng, Yang, & Yen, 2017). In the grow-out system, this means that a large number of females can still reproduce and produce a large number of fries. There is a lot of material on the YY-male technological approach, which is meant to create all male fry by crossing YY-males with any females in theory. When numerous farmers have tried this strategy, however, they have discovered that the outcomes are inconsistent and unpredictable. As a result, they see dangers in commercializing the approach. The most extensively used standard technology for making all males is hormonal sex reversal. Hundreds of private hatcheries in Bangladesh, Brazil, Thailand, and Vietnam, as well as several African countries such as Ghana, Malawi, and Zambia, have adopted the technology, producing and supplying billions of high-quality monosex tilapia fry each year.

One of the most profitable rural companies is the construction of a specialist tilapia hatchery and supply of high-quality monosex fry. The net profit from the tilapia hatchery industry has been highly competitive when compared to many other industries. As a result, many poultry, carp, prawn, and other hatchery operators have either totally converted to a tilapia hatchery company or at the very least accepted it to diversify their risk. The management of brood stock in hapas, artificial incubation of eggs in downwelling incubator jars, larval rearing in shallow trays, and the hormonal sex reversal technique as it has significant benefits and there is strong interest in its use around the world are all part of the business of mass-scale tilapia fry production using hormones. The technique was developed in Thailand, where a series of research trials were carried out over the past three decades, funded by various organizations. Hundreds of hatcheries popped up in Thailand, and they quickly spread to Bangladesh, where there are now over 300 hatcheries. However, due to the rapidly expanding demand for tilapia, the demand for fry remains unfulfilled. A single hatchery may produce and serve over 30 million fry each month with this technology. Planning and creating goals are the foundations of any business. Volume targets, number of specific product targets, and net profit targets are all possible. The targets in this situation are the quantity of fries to be sold and the amount of profit to be made. Targets may differ depending on the size of the company, the demand for the product, and the company's goals. Small, medium, and large size tilapia hatcheries are the three types of tilapia hatcheries. The cost–benefit analysis is based on anticipated item costs in Asia. Capital costs have been included as depreciation costs, which have been computed based on the fixed assets' lifespan. This type of hatchery is becoming increasingly popular, owing to its high profit margins. Some of the early hatcheries in some nations took advantage of high demand and made large net profits, sometimes exceeding 100%(Yap, Villaluz, Soriano, & Santos, n.d.).

When choosing a location for the hatchery, (Figure 1.3) consider the local demand for tilapia seed as well as the possible profit from the enterprise. Hatcheries and nurseries are generally clustered together. This follows a marketing trend in which certain locations may already have a good name for certain products. As a result, that location may be selected over a hatchery in another location. There is a risk of rivalry in clusters with easy access, thus fry quality must be maintained at a much higher level. The disadvantages of living in a remote region are evident. However, if the hatchery has an access road, seed, broodfish, and other hatchery materials can be moved to and from the hatchery in a reasonable amount of time and at a low cost. Although fry can be packaged in plastic bags with oxygen and sent great distances i.e., up to a 20-hour drive or 40-hour shipping involving air cargo and ground transportation throughout the world, distance from grow-out growers is important ("Transportation of Fish in Bags," n.d.). Long excursions, on the other hand, can deplete fry and jeopardize their survival. There should be ample room for brood stock, incubation/larval raising, and fry nursing and holding in the pond, hapa, or tank. The quantity and size of ponds, hapas, cages, or tanks, on the other hand, is determined by the size of the operation. The most popular type of hatchery is one that combines a hapa with a tank. Hapas are simple to set up and less expensive. They are also beneficial in terms of maintaining brood stock purity. Most other species' hatcheries are tank-based and often extensively equipped with recirculation systems, aeration, and light management, among other things, which is why hatchery operations are considered high-investment and high-tech businesses. However, this is not the case with tilapia, and hatcheries have sprung up everywhere, even in the world's poorest countries.

Figure 1.3: Tilapia hatchery.

Source: https://i2.wp.com/fishconsult.org/wp-content/uploads/2011/11/Commercial-tilapia-hatchery.jpg

1.4 STOCKING

Fish stock management is done in pond-based systems in most tropical and subtropical nations since it is the most cost-effective way. However, because wild fish and other predator organisms can easily enter ponds, the survival and purity of brood stock and fry/fingerlings cannot be assured. Installing hapas in ponds, on the other hand, can make it much easier to manage fish and preserve a pure population. Because they make it easier to control or handle the fish, are cost-effective, and enable water exchange from the pond, hapas are extensively employed in ponds to maintain brood stock, as well as for fry nursing and sex reversal. Hapas range in size from $5m^2$ to $120m^2$ depending on the purpose ("Carp Cultivation and the Highlands Aquaculture Development Centre, Aiyura," n.d.). A hapa (Figure 1.4) is a suspended structure made of bamboo sticks strung together with nylon ropes or concrete pillars tied together with metal strings. The hapas should be 20–40 cm above the pond's floor. A hapa's water depth should be kept at 60–70 cm, and its walls should be raised 20–30 cm above the water surface to prevent fish from swimming away(Apud, Primavera, & Torres, 1983). A hapa should not cover more than 40% of the surface of the water. In many situations, hatchery operators install too many hapas, resulting in water quality issues, which can lead to fish death, low egg production, fry mortality, and other issues.

Figure 1.4: Hapas.

Source: https://www.researchgate.net/profile/M-G-Hussain/publication/283348879/figure/fig19/AS:645371718025227@1530880016489/Tilapia-fry-holding-hapas_W640.jpg

Underground tube wells, irrigation canals, streams, rivers, reservoirs, and thermal plants are all possible water sources for fish farms. Because water is scarce in most regions, a water recirculation system for artificial egg incubation is an important feature of any tilapia hatchery system. Water availability is one of the most crucial aspects to consider when choosing a location for a hatchery, as it must be available 24 hours a day, seven days a week. Water for hatcheries should be sufficient, inexpensive, and devoid of microbes and contaminants. The water temperature in a hatchery should be between 25 and 35 degrees Celsius, with 28 to 32 degrees Celsius being ideal(Nobrega, Banze, Batista, & Fracalossi, 2020). Most hatcheries use pumps to bring underground water up to the header tank, usually from very deep tube wells that reach up to 100 m. While some hatcheries are fortunate enough to obtain clear and well oxygenated water from streams and rivers by simple diversion, most hatcheries use pumps to bring underground water up to the header tank. DO levels in underground water are low, and green water can have DO levels as low as 0 mg/l as early as 6 a.m. Although tilapia can survive for brief periods of time in these settings, DO should not be less than 4 mg/ l (Makori, Abuom, Kapiyo, Anyona, & Dida, 2017). Before pouring into the pond or tanks, a simple splash can be generated at the tip of inlet pipes to oxygenate the water.

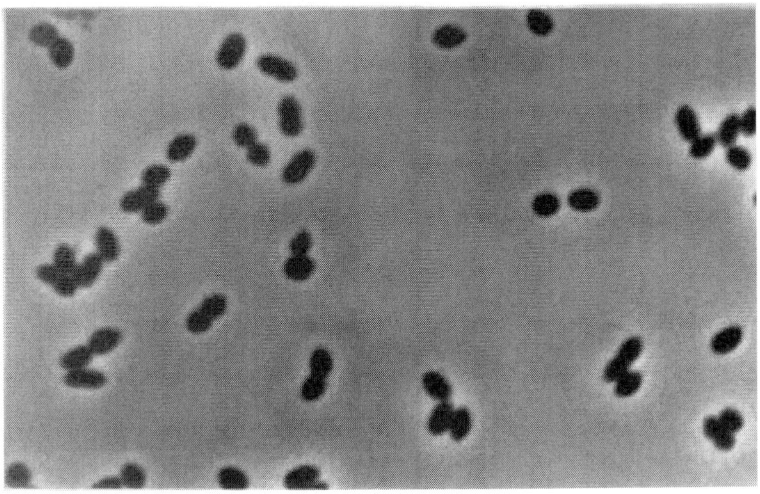

Figure 1.5: Nitrosomonas bacteria.

Source: https://upload.wikimedia.org/wikipedia/commons/2/24/Nm-eutropha.jpg

Nitrosomonas bacteria (Figure 1.5) can consume dissolved nitrogen and convert it from ammonia to nitrite and nitrate in some plants. The sand filtration system, on the other hand, separates coarse solid particles from water and filters or stops some organisms from entering the system. Perforated pipes should be installed beneath the sand filtration system's bottom so that enough filtered water may be stored for pumping and lifting to the header tank. UV light has been found to be the most cost-effective method of sterilizing water, particularly for egg incubation systems. Just before it is supplied, water can be passed through a simple UV light tube installed inside a 4-inch PVC pipe. Because UV light kills all germs, it can be harmful to eggs and larvae if the intensity is too high, so agents who supply UV lights should be consulted for specifications. Ozone kills all germs and has been found to be effective in water sterilization. However, because the amounts of bacteria and water are so large, it is relatively expensive and may not last long. As a result, more research is required to make it appropriate. Running a fish hatchery does not require a lot of electricity, but it is necessary because egg incubation systems, hatching, and larval rearing in aluminum trays require a constant supply of filtered and clean water to churn them and provide oxygen. If the incubation and larval rearing are done in the same recirculation system, a basic 3 hp pump should suffice. Some hatcheries require two pumps to maintain these two systems distinct so that if one fails, the hatchery may continue to work without interruption. Because fry concentrations are high during these stages and fry mortality is possible, an additional air pump may be utilized in most hatcheries to supply aeration in conditioning tanks and any sex reversal system.

A basic aeration system consisting of an air pump and numerous air stones connected by small rubber pipes can dramatically enhance survival. Because it will form the foundation of the stock for a long time, the first brood stock should come from a reputable source. The domesticated stock should be obtained from well-established hatcheries, even though it can be collected from natural or feral herds. Then, with correct management, hatchery operators can generate their own brood stock on-farm, which is more reliable and less expensive. Furthermore, they will have complete control over the stock. However, maintaining purity and avoiding inbreeding is vital, and this can necessitate a large amount of knowledge, expertise, and resources. To avoid inbreeding, a new hatchery should start with at least 10,000 broodfish with an expected sex ratio of 1:1(Hussain, 2004). Medium-sized hatcheries may purchase 10,000–20,000 brood fry, while large-scale hatcheries may purchase as many as 100,000, allowing them to begin breeding and fry production right away without having to wait for another batch.

Swim-up fry gathered from a tray system can be put in seven foam boxes if the brood fry demand is 25,000. Each carton has four plastic bags, each containing about 900 swim-up fries in approximately 2 liters of water and 3 liters of oxygen. If the brood fry is purchased from outside the country, a health certificate based on samples tested within a week of departure is required. Normally, the government agency in the exporting country issues a certificate. It's also possible that a certificate of origin is required. Some countries prohibit the importation of exotic living animals, while others may provide import licenses subject to severe laws and regulations, including a lengthy quarantine and surveillance process. Fries that have been adequately conditioned before packing, meaning they have been kept in clear water for at least 24 hours, can survive up to 36 hours of travel. Brood fry can therefore be imported from anywhere in the world. Tilapia can be bred from the age of six months, or when females weigh around 100 g and males weigh around 150 g. Fish with good shape and appealing color should be chosen from the maturation pond or tank or hapa; any malformed fish or those that are larger or smaller than usual should be avoided. The dorsal and caudal fins of males with a pink coloration are favored. Broodfish can be employed until they reach the age of two to three years, depending on their size. Smaller females will produce fewer eggs, resulting in fewer fry. As a result, some hatcheries stock larger females weighing around 150 g, together with 200 g males, to stock in breeding hapas(Mohammed et al., 2014). Females and males weighing more than 350 g and 400 g are often difficult to manage during seed harvesting; as a result, they are isolated and sent to tanks called hapas, where they will be sold in a month or two for a profit.

In natural breeding, each sexually active male tilapia constructs a nest, similar to how birds do. A mature female with well-developed eggs will follow a nest if it sees one. The males and females encircle the nest, indicating that they are breeding. The female then lays eggs, and the male fertilizes them with sperm. The same female picks up the eggs or embryos in the mouth after fertilization to incubate them. It keeps them in its mouth for up to two weeks, or until they're fully capable of swimming freely and surviving on their own. Even when the fry is swimming freely, the mother remains nearby. It extends its mouth to provide a safe space for the free-swimming fry if it sees other fish or predators approaching to attack them. This is an inherent feature of Nile tilapia maternal mouth brooding, and it resembles the traits of terrestrial animals at a later evolutionary stage.

When females lay eggs in mud nests, it's difficult to pick up the fertilized eggs without picking up mud and detritus as well, and some of the eggs may end up lost in the nest. As a result, Hapa systems are superior since all of the eggs may be gathered. If the water quality is inadequate, however, hapas can get contaminated, and dirty hapas should be replaced every 2–3 months after an egg harvest. Furthermore, females that hold embryos and yolk-sac fry in their lips do not feed, resulting in weight loss. They strive to defend their young and themselves as much as possible, but when they are attacked by other fish or predators, as well as when environmental conditions or water quality make life difficult, they may lose some of the embryos and yolk-sac fry. Because the water quality in most ponds isn't ideal for these delicate embryos, collecting embryos as soon as they're placed and fertilized and transferring them to an artificial system with clean water for incubation will boost productivity. Seed refers to the eggs, embryos, yolk-sac fry, and swim-up fry that have been gathered. Fecundity is the primary indicator or measure of seed productivity in most fish. Fecundity does not immediately correspond with seed productivity in tilapia since some of the parameters listed above affect the quantity of eggs or embryos that can be gathered. Mouth-breeding tilapia females incubate eggs in their mouths until the young are able to swim independently. Long scoop nets can be used to gather these free-swimming fries from the boundaries of the pond, hapa, or tanks at intervals of 7–21 days. This system is inexpensive and simple, but it produces relatively little net fry per unit area of space because it is hard to harvest all of the fry from the system. As a result, it's also known as the partial harvesting method. Predation and harsh weather conditions also make survival difficult. Furthermore, fry vary in size and age, and partial harvesting does not give excellent results when using sex reversal techniques. Harvesting more regularly, such as two or three times a day, will result in a more uniform age and size of fry. Grading fry is critical in partial harvesting and can be done with a variety of mesh sizes.

Figure 1.6: Fry.

Source: https://assets-api.kathmandupost.com/thumb.php?src=https://assets-cdn.kathmandupost.com/uploads/source/news/2017/others/p3-lead-19062017083254.jpg&w=900&height=601

Fry can (Figure 1.6) be nursed for up to a month in hapas, ponds, canals, tanks, and other similar environments. Several attempts have been made to produce completely male tilapia fry, the hormonal sex reversal procedure has established to be the most reliable and popular. The sex reassignment process is separated into two phases: a 21-day hormone therapy phase and a 10–30-day nursing period. When young fry is given a male hormone like 17-alpha-methyltestosterone, they become fully masculine phenotypically(Little, Bhujel, & Pham, 2003). This prevents breeding in grow-out systems, diverting energy that would have otherwise been wasted on gamete production and reproductive activity. 60 mg of 17-MT is mixed with 1 kg of fishmeal and 20 g of vitamin C or vitamin combination for sex reversal. The swim-up fry is then placed in a sex reversal system like hapa in ponds or tanks. The fry has not been fed up to this point. Fries are 10–11 days post-hatching when they are transplanted and MT feeding begins. Because sex determination occurs 12–15 days after hatching, it is critical to begin MT feeding at this age (Chiasson & Benfey, 2007). When they begin eating MT feed, the level of testosterone in their bodies rises, causing the sex to shift toward male. Sex reversal will be impossible if MT feeding is started late, or when they are older than 15 days. If hatchery operators want to achieve the goal of generating all or virtually all male fry, this is one of the essential concepts they must remember.

Several trials have been carried out in order to improve survival and increase the proportion of males in fry populations. These included determining the optimal MT dose in feed, as well as the frequency and length of the feeding interval. As a result, a consistently high percentage of men has been obtained. Nursing and advanced nursing methods have also been developed through study when fry must be kept for a longer period of time. Keeping records is a time-consuming task, but it allows management to assess the current condition and change it if something goes wrong, avoiding a major loss. More importantly, analyzing time series data and recognizing simple trends can assist in steering the manufacturing line depending on seasonal demand and sales variations. There are always valuable lessons to be learnt, and keeping such data has become increasingly important in the context of Best Management Practices (BMP) and certification. A tank system with water recirculation can likewise be used for sex reversal. Some hatcheries feed hormones for 7 days while maintaining fry in tanks, then move them to a hapa-in-pond system for 14 days or keep them in the tank for the entire 21 days(M. S. Ali, Griffiths, Turner, & Farms, n.d.). When compared to a green water hapa-in-pond system where plankton is abundant, feeding fry in tanks makes them more likely to consume hormone-containing feed. In tanks, extra caution should be exercised because fry will be stocked at high densities, and providing high-protein feed will result in excessive ammonia production. There is a limited amount of space available, and there may be a high rate of death. Aeration is required, which comes at an additional cost to the farmers.

Fry nursing can be run as a separate and self-contained enterprise. It takes little time, so nursing farmers can earn more quickly and on a more consistent basis, possibly monthly. In most situations, however, the hatcheries are responsible for the fry's care. Grow-out farmers, on the other hand, may want to nurse fry to ensure that they have enough fry when they need them. Smaller fry can be more cost-effective for them because the price of small fry is low. Smaller fry, those weighing less than 5 grams, are more vulnerable to predation by other fish, particularly carnivorous species like catfish and snakehead, as well as snakes, birds, and frogs. Smaller fish are also more susceptible to poor water quality, which is frequent in ponds with organic manure. These factors raise mortality rates, especially in the first month of growth. As a result, if they are nursed for at least a month before being placed in grow-out ponds, their chances of survival are much improved. Tilapia may be able to withstand harsh environments than other animals, such as high salinity and cold temperatures. Because Nile tilapia can only be cultured in regions where the salinity is less than 10 parts per million, it has some limits. It cannot be cultured in coastal or brackish water. Acclimatization of fry during conditioning, on the other hand, may aid to some extent. Before being transferred to these systems, Nile and blue tilapia fry must be gradually acclimated to brackish water. The salinity of the water in the conditioning tank should be raised by 2–3 ppt per day until it reaches the same level as the grow-out farm where they will be stocked. Red tilapia fry adapts to brackish water more quickly and can thrive in higher salinity environments. Tilapia is a tropical fish that grows slowly at water temperatures below 20°C(Rahman, Shahjahan, & Ahmed, 2021). This is why tilapia cannot be grown in northern hemisphere areas such as Europe, North America, and northern Asia unless they use heated water from thermal plants or other sources. They must rely on countries in the tropics and subtropics. They might perform well if the fry is gradually acclimated to lower temperatures.

During conditioning and packing, between 5–10% of the fry may be lost. When there is a power outage, bumpy roads or shaking of the fry, hot weather, or any other event that creates poor conditions like low temperature, low pH, low DO, or physical stress, large mortality rates can occur. During the day, especially during hot and dry seasons, fry transportation should be avoided. Plastic bags should be covered with thick sacks if necessary. To keep temperatures down before and during transit, water can be sprinkled over the bags. Fry can be transported via any kind of transportation. In Bangladesh and Vietnam, a basic bicycle or motorbike can be used to transport fry over small distances. By road, cars, pickup trucks, and trucks can transport fry for a full day. Fry can be sent to any location on the planet. Each foam box, which should be clearly labeled with the recipient's name and correct address, usually has four bags containing at least 2000 fry. A health certificate provided by a government body is required for cross-border shipment to ensure that the fry is clear of parasites and disease-causing pathogens. Importing farmers, on the other hand, may be required to get a license and adhere to quarantine requirements.

1.5 TRANSPORTATION

Plastic or steel tanks with an aeration system can also be used to transport fry or fingerlings. Farmers in Bangladesh use aluminum or earthen vessels that they carry on their bicycles or on their shoulders, hanging two vessels from each end of a bamboo staff. During transportation, they manually agitate the water with their palms to introduce oxygen from the atmosphere. When releasing fry into the culture system, allow the plastic bags to float in the reception tank or pond for a few minutes so the fry may gradually adjust to the new water temperature. The tilapia hatchery business has been extremely popular in Asia, indicating that the technology that has been developed and distributed has been extremely successful. In other words, despite its comparatively costly investment, it has proven to be commercially viable. According to this estimate, the gross and net profit margins are both over 75% of the investment. Feed for broodfish is a considerable expenditure, accounting for more than a quarter of total costs, followed by personnel and MT hormone. The total variable cost is 65 percent, while the remaining fixed costs are the rest. Marketing and transportation of fry are the largest fixed costs, followed by energy and fuel. Earth ponds, floating cages, and concrete or plastic tanks are some of the existing methods used in most countries. The cultivation of tilapia in ponds spans from small-scale subsistence to large-scale commercial agriculture. Ponds can range in size from 100 m^2 to tens of thousands of acres. The most common culture system is ponds, followed by cages. Tilapia can also be raised in tanks and raceways, although this is too expensive for small-scale farmers.

Tilapia have a variety of eating patterns. Most tilapia eat zooplankton, debris, aquatic plants, insects, and even small fish fry in addition to phytoplankton. Commercial pellets, discarded food, and practically any other sort of feed are consumed, with the exception of meat. If they are cultured in green water ponds, very little investment in their nutrition is necessary. With increased stocking density, production costs will rise. Only if the stocking density exceeds 2 fish/m^2 will feed be necessary. When fish are stocked in larger densities, the type of feed used is dictated by the price of the fish. If the price of tilapia is very low, as it is in most of Asia, feeding huge amounts of commercial floating pellets will not be cost-effective. In such a case, only low-cost feed inputs will be cost-effective. Maintaining a green pond is the most critical aspect of tilapia rearing. The more natural food that is available in the pond, the faster the fish will grow. Too much green, on the other hand, can be harmful. A Secchi disc is used to determine greenness. A drop in visibility of the Secchi disc between 20 and 30 cm should suggest optimal transparency/turbidity of the ponds. Low oxygen levels early in the morning might induce fish mortality if a pond is too green. Fertilization using chemical fertilizers and/or animal dung regulates pond greenness. The pond will become greener as additional fertilizer is introduced. Water color should be used to determine the amount of fertilizer required. Increase the amount of fertilizer if the pond isn't very green. Stop fertilizing if the pond is too green in the morning and the fish are gulping. During the growing period, the amount of fertilizer/manure required will generally rise.

Chemical fertilizer should be applied by dissolving it in water and disseminating the solution across the pond's surface from at least two places, depending on the pond's size. The application of animal manure is not governed by any rigid rules. The majority of farmers either distribute the manure or apply it to a few specific areas around the pond's perimeter. Manuring in little amounts on a regular basis is recommended, but once a week may be enough. Manure is placed in bags and hung on ropes by certain farmers. Increased levels of feeding and stocking densities can result in larger productivity and total volume of fish produced, but this does not always imply a higher net profit. Farmers may even lose money if they are not diligent and do not calculate correctly due to excessive levels of investment. This is the main issue among farmers, as the majority of them do not keep records and are unable to calculate cost–benefit ratios. A combination of fertilization and supplemental feeding is likely to be the most profitable. Simple economic models with extensive financial information must be developed in order to assist farmers or attract possible investors to the aquaculture industry. They need to know how much money they'll have to invest and how much money they'll make from different scales of aquaculture systems.

1.6 MANAGEMENT

Management systems, material and equipment availability, and costs vary greatly based on local conditions. As a result, models must be adjusted to fit the circumstances. There is a good profit based on the economic model established for the Bangladeshi situation. Feed is the most expensive component, accounting for 65 percent of the entire cost. The profit, on the other hand, is the cost of management, which is done by the farmer himself. If the farmer operates the farm on his or her own, he or she can earn about $7900 per year, or $658 per month (Tran et al., 2021). This is enough to entice a business owner to begin with. The location should have enough water current to provide DO to the fish while also removing metabolites from the area around and beneath the cages. Temperature, oxygen, pH, turbidity, and other variables are critical for cage culture. Water should have a minimum DO concentration of more than 3 mg/l. Tilapia growth is hampered by DO levels below this level. For tilapia cage cultivation, the optimal temperature and pH ranges are 22–32°C and 6.5–8.5, respectively. Strong winds and waves should be avoided because they might move cages or even damage them, allowing fish to escape. Because most natural water bodies are located far from residential areas, protecting them from poaching is a serious issue. Barbed wire and an alarm system should be used to secure the area or access road. Each cage should have a lockable net on top if at all practicable. Even when nets are rocked and shifted by powerful currents or waves, fish would be protected. The cage area should be guarded 24 hours a day by either a fence or security guards. Poaching will be deterred by good lighting in the region, which will also make harvesting easier. Water flow may be hampered by arranging cages in a rectangular compact manner. Cages are set across rivers to ensure that all cages receive fresh water. If the river is to be used for transportation, however, cages cannot be placed over the river; instead, they must be placed along the river. Cages can be arranged in a variety of ways in lakes and reservoirs, depending on the water current and its direction, as well as anticipated winds, and often just depends on the farmer's preference. Following a huge loss of fish due to river water contamination, some farmers are switching from cage culture in rivers to cage culture in ponds.

Communal water bodies are utilized for a variety of purposes, including the transportation of goods such as rice, cement, molasses, and other commodities, and enormous barges have been known to capsize. Pesticides pouring from agricultural areas upstream have also contaminated river water in the past. Some farmers construct cages in their own ponds to prevent these issues. Farmers keep large fingerlings and young adults, weighing 50–100 g, in cages and feed them high-quality pellets at a high percentage of biomass or close to satiation to ensure quick growth. Farmers can shorten the grow-out period to 3–4 months and harvest fish weighing 600 g to 1 kg, which appear fat and command high prices. Another advantage is that the pond's fish do not require feeding. Unused feed and nutrients leached into the water are devoured directly by fish or by plankton, which are then ingested by the pond's fish. Tilapia in cages is actively maintained by supplying entire commercial pellets, but they can also be cultivated in a semi-intensive manner. In a semi-intensive system with low-density culture, rice bran and fishmeal can be mixed together as a home-made feed and fed to tilapia. Pellets containing 25–35 percent CP, both floating and sinking, are employed. Feeding rings are commonly used to keep floating pellets contained within the cage. Feed should be supplied on a feeding tray buried below the water level since sinking pellets dissolve quickly in the water. Feeding can occur once or twice a day, in the morning and late afternoon. Farmers often feed tilapia 3–5% of their body weight every day. Higher feeding rates, on the other hand, are typical among those who expect quick growth and larger tilapia harvests, such as in Thailand, where farmers are instructed to feed 5–8% biomass daily so that they can collect large fish that appear to have more meat.

Depending on feeding and other care, the grow-out culture duration might range from 3 to 7 months. If the fish are seeded at 40–50 g, it may take four months for them to grow to 750 g. However, aquaculture is getting more sophisticated, and farmers now stock approximately 100 g of huge fish, which may grow to 750 g in three months, with cage survival rates of 70–80 percent. In the Philippines, a 100 m2 cage produces an average of 540 kg of meat. Cages are frequently found in rivers or lakes that are below the road's level. Farmers in Thailand use pulleys to hoist the fish from the cage while harvesting with a motorcycle engine. After lifting the fish to the roadside, they weigh them and then load them into a pickup truck. Farmers, on the other hand, appear to be satisfied because it is a time-saving task, especially because when cages are constructed, housewives can take care of feeding and other minor management, allowing them to avoid working in the fields as agricultural labor or in other low-paying employment. It has produced jobs for people who would otherwise be unemployed or working as laborers in rural areas. It is achievable in a country where either the government or commercial firms provide excellent extension services, supply fingerlings and feed, and then buy back the fish for processing and marketing. However, the amount of fish that farmers may collect has a significant impact on revenue and profit. Net profit is only achievable when fish weigh more than 400 grams, and if a large percentage of the fish are small, farmers will lose money. As a result, the quality of fingerlings is critical to the success of cage culture. Farmers in Asia hunt for monosex fingerlings to stock in their cages for this reason. If 20% of females do not reach a weight of 400 grams, the revenue may be reduced, resulting in net losses.

This will need to be thoroughly examined in other nations to determine if it will work - it may not work in areas where such services are lacking, and even if they do exist, the quality of services, fingerlings, and feed can be an issue in many circumstances. In such instances, vertically integrated corporate farming is required to ensure that fingerlings are produced by the same farm and that fish may be easily marketed, either locally or internationally. With the exception of a handful, this is the scenario in most African countries. As a result, only major corporations have been able to commercially raise tilapia. Cage culture, for example, is currently regarded as an ideal vocation for landless families who live near water bodies but live temporarily on others' land or communal land, and who struggle to make ends meet on salaries earned as laborers. Tilapia cage cultivation can generate a decent income without the necessity for land ownership. The government has granted them permission to use the water because it is considered a community welfare activity. The government and other community development organizations in this case prioritize the creation of jobs, the production of fish for family nutrition, and some income from surplus fish, rather than maximizing production volumes and profits, which would be the priority if cage culture were carried out by the private sector.

CHAPTER 2

TILAPIA FARMING

CONTENTS

2.1 INTRODUCTION

Tilapia is Egypt's most important aquatic species, both in terms of production and consumption. Tilapia production has increased over time, reaching 687,400 Mt in 2010, valued at US$991 million, accounting for 38% of the overall fish value and 53% of total fish output(Nahid, Karim, Hasan, Shahabuddin, & Bhuyain, 2012). The output of tilapia pond culture is a function of a variety of variable and constant inputs employed during the manufacturing process. A crucial aspect of a successful culture operation is determining the best number of inputs to employ. The optimal output level is determined by requirements, which define how much of the input to use. As a function of input and output pricing, input demand equations show how much of each input will be purchased and consumed. These functions are derived from the profit maximization first-order condition (FOC). The difference between total returns and total costs is used as the profit function. We get the profit maximizing demand functions for inputs by setting the first partial derivatives of the profit function with respect to input prices to zero and simultaneously solving for input quantities as a function of input and output costs. These demand functions reveal the exact input levels that the firm uses to keep each input's VMP equal to its purchase price. Because of the complementary relationships between feed and fingerlings, rising feed prices combined with relatively low fish selling prices will reduce demand for fish feed and fingerlings. This will have a significant impact on the industry's long-term viability. In the United States, for example, recent increases in catfish feed prices and relatively low fish selling prices have resulted in industry shrinkage since 2003. In tilapia pond cultivation in El-Fayum, decreasing returns to scale appear to exist. Producers were found to be overstocking their operations by 36 percent to 115 percent and underfeeding by 8 percent to 22 percent. The optimal stocking weight and rate, feeding rate, yield, and profits will be reduced if feed prices rise or tilapia selling prices fall. A proportional change in feed price has a bigger impact on stocking and feeding rates, yield, and profits than a change in tilapia selling price. Local feed prices climbed by around 57 percent between 2008 and currently, while tilapia prices increased by 16 percent to 18 percent(Kassam, n.d.). As a result, farm-level efforts to reduce feed costs and improve feeding methods are required.

2.2 DEMAND

According to recent predictions from the United Nations' Food and Agriculture Organization (FAO), the world's population would rise from 7 billion to 8.3 billion by 2030, and 9.1 billion by 2050, requiring a 60 percent increase in the production of food over the following 30 years("World Population Projected to Reach 9.8 Billion in 2050, and 11.2 Billion in 2100 | UN DESA | United Nations Department of Economic and Social Affairs," n.d.). Furthermore, global middle-class per capita income is predicted to rise, and rural-urban migration is expected to level off in the next 20 years. Fish is one of the healthiest proteins, given its amino acid and fatty acid composition, and rising affluence is predicted to push preferences toward healthier cuisine. Approaches to fulfilling growing food production demands will most certainly differ by country, and country-specific analyses of regional supply and demand can be useful policy inputs. In the last three decades, global aquaculture production has expanded by over 1,100% at an annual rate of 8.8%(Naylor et al., 2021). In 2016, global fisheries production was 170 million metric tonnes, making fish the most abundant source of animal protein(Hua et al., 2019). In the previous ten years, the capture fisheries have stabilized at around 90 million metric tonnes(Ritchie & Roser, 2021). For the first time in 2011, aquaculture produced more protein than beef, and in 2013, consumers devoured more fish from aquaculture than from catch fisheries worldwide(Ben Belton & Thilsted, 2014). Brazil, like many other countries, is investing in and upgrading its aquaculture supply chain. Official figures from the Brazilian Institute of Geography and Statistics show a robust development trend in aquaculture in Brazil, with production increasing by 133 percent in farmed fish between 2009 and 2013 and reaching 547,163 metric tonnes of fish in 2017(Valenti, Barros, Moraes-Valenti, Bueno, & Cavalli, 2021).

Figure 2.1: Brazilian aquaculture.

Source: https://aquavitaeproject.eu/wp-content/uploads/2021/03/mangrove_oyster_production_in-_cananeia_web.jpg

Other factors, such as the availability of water resources, have aided the rise of Brazilian aquaculture (Figure 2.1). Aquaculture facilities in freshwater lakes of hydropower plants have the potential to expand, however they are currently underutilized. There are 219 freshwater reservoirs in 22 states, spanning 3.14 million hectares of land(Marques, Watterson, da Rocha, & Cavalli, 2020). Brazil has high temperatures throughout the year in some areas, and there is a vast range of native fish and shellfish species, including more than 60 farmed kinds. There is a sizable consumer seafood market, scientific expertise to assist production, and abundant grain production, mostly corn and soybean, which are key ingredients in fish feed. Aquaponics is the combined production of fish via aquaculture and plants via hydroponics (Figure 2.2) in an integrated growing system, in which waste from fish production is transformed into beneficial nutrients for plants by bacteria.

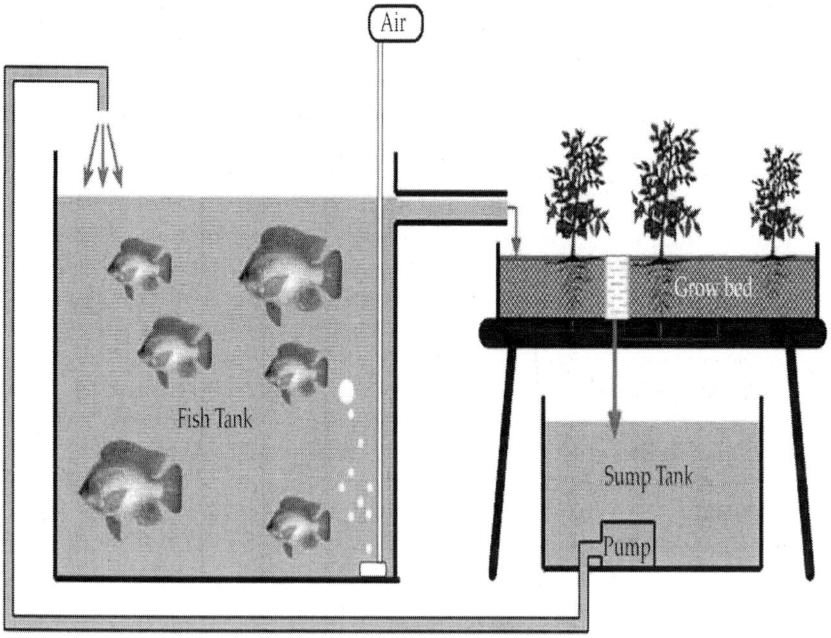

Figure 2.2: Hydroponics.

Source: https://aquapona.co.uk/wp-content/uploads/2019/02/what-is-aqua-ponics-diagram-2-1024x683.jpg

The interlinking of aquacultural and hydroponic methods allows some of the inadequacies of the respective systems to be solved and thus constitutes a viable sustainable food production technique. Vegetable production in an aquaponics system might be termed organic using this strategy of fertilization, and the higher prices generally obtained for organic vegetables could provide economic incentives for commercial aquaponics production. Most commercial aquaponics production is now done in greenhouses or indoors, where the climate is regulated, using a combination of hydroponics and aquaculture techniques and instruments. Researchers are attempting to understand the economic feasibility of aquaponics production, which continues to pique attention around the world. It was found that tilapia is the species of fish grown by the greatest number of producers in a survey of aquaponics businesses in the United States. According to the survey, just about a third of those businesses are profitable. Starting with catfish and learning the system's administration before switching to more expensive species like tilapia for specific markets is a solid method for producers to have success in an aquaponics operation. Other studies have looked at the profitability of aquaponics enterprises in various regions.

2.3 LEGISLATION

Tilapia is the most commonly farmed fish in Brazil, accounting for over 58 percent of total farmed fish production by weight(Cai, Zhou, Yan, Lucente, & Lagana, 2019). However, tilapia farming is forbidden by legislation in some parts of the country, while other native Brazilian species such as tambaqui and pacu have high potential due to market acceptability in Brazil. The scientists discovered a link between faster water recirculation and improved fish survival rates, but they did not calculate the economic benefits. Some show that pacu is an economically viable alternative by using red tilapia and pacu effluents in aquaponics production with onion and parsley. Weighing the fish every day was impossible because it required capture, anesthetic, weighing, and returning the fish to the tank, all of which stressed the fish and may have hampered its subsequent food intake and growth. The fish were raised until they were juveniles. These juveniles, also known as giant fingerlings, have a high market potential as commercial fish grow-out inputs and can be caught within the 30-day production period(Ragasa et al., 2022). There is a good market for juveniles since it is more cost-effective for fish farmers to buy juveniles rather than young fingerlings in order to optimize the production structure and minimize the first losses in grow-out due to predation in ponds and reservoirs. The economic benefits of developing grow out fish from larger juveniles have been demonstrated. Certain floating feed was found in some treatments with feed rates above the recommended fish stocking density, indicating feed wastage. When the feed waste was discovered, it was removed from the tank since it was lowering the water quality and posing a risk of damaging or clogging the aquaponic system.

Figure 2.3: Nile tilapia.

Source: https://upload.wikimedia.org/wikipedia/commons/3/3f/Til%C3%A1pia_ou_Sarotherodon_niloticus_2.jpg

The Nile tilapia (Figure 2.3) is the most widely farmed freshwater tilapia species, accounting for over 71% of global tilapia production(Modadugu & Acosta, 2004). Fast growth, rapid food conversion, high fecundity, tolerance to a wide range of environmental factors, and good table quality are among the additional characteristics of *O. niloticus*. Because of a lack of economic data in most African nations, fish farmers find it difficult to obtain financial support from the government or financial institutions. Economic data such as production costs and sales could help fish farmers figure out how profitable their operations are. This information will help fish farmers get the financial help and support they need to invest and develop their businesses. The general purpose of research is to calculate the cost of growing *O. niloticus* in a semi-flow through culture system with two different feed types. Integrated fish farming involves the production of fish, livestock, and crops in the same production unit, with a linkage and synergy between the various products. In several countries, integrated aquaculture has been used to boost fish productivity by fertilizing fish ponds with a variety of organic fertilizers such as poultry, cattle, and pig manure. In freshwater integrated agro-aquaculture, poultry manure is commonly used. Integrated agro-aquaculture has the potential to provide not only a more equitable distribution of income-generating opportunities but also a more efficient and environmentally sustainable utilization of scarce resources. The integration of fish with animals and crops has a lot of potential for increasing animal protein production, creating jobs, and improving farmers' socio-economic conditions. Integrated livestock-fish-crops have been proven to be not only technically but also commercially successful.

In addition to enhancing total production, income, and employment, integrated farming methods improve environmental sustainability by recycling waste and lowering the risk of pollution. In an integrated farming system, output from one subsystem that would otherwise be wasted becomes an input to another subsystem, resulting in a higher efficiency of output of desired product from the land/water in the farm. Incorporating fish farming into an integrated system maximizes the use of resources such as water, farm by-products, land, and labor. Pond water, for example, is used to irrigate domestic crops and provide water for animals in addition to farmed fish. Pond water is richer in nutrients than well water as an irrigation source, and it also contains nitrogen-fixing blue-green algae, which can boost soil fertility. While some crops, such as vegetables, can be directly watered from pond water, others, such as bananas, sugar cane, and yams, can benefit from pond moisture. In Tanzania, integrating fish farming technology into the existing farming system has been seen as a viable approach for enhancing agricultural production. Furthermore, phosphorus, nitrogen, and potassium are necessary components for plant growth in the soil. The soil will be rich in those three components from the pond water fish pond during the flushing of pond water into the garden, increasing agricultural productivity surrounding the fish ponds. Because of the multiple outputs in the system, integrated agro-aquaculture enables economic growth. Instead of a single product, chicken, fish, and vegetable production could contribute significantly to wealth development in a production unit. The elimination of the expense of water for vegetable irrigation is another economic benefit of integrated agro-aquaculture. Integrated aquaculture minimizes the farm's production expenses. The goal of on-farm integrated agro-aquaculture (Figure 2.4) is to optimize profit from the integrated production unit by allocating resources and managing by-products. Sub-systems can be integrated to make use of end products, by-products, and trash.

Figure 2.4: Agro-aquaculture.

Source: https://onlinelibrary.wiley.com/cms/asset/ff8c1d6e-0c0e-4e37-84b6-84fb8d499ebf/jwas12714-fig-0005-m.jpg

2.4 PRODUCTION SYSTEMS

The current aquaculture production system in East Africa is small-scale clay ponds with low inputs and yields(Tidwell & Bright, 2019). In Malawi, researchers evaluated the social and economic impact of integrated agro-aquaculture, and in Uganda. Nutrition, food security, and livelihoods are all impacted by aquaculture. In human nutrition, fish is a major source of dietary animal protein. Aquaculture development of aquatic species for protein supply has recently been encouraged all over the world. Fish is a suitable protein substitute for red meat in the diet. Fish has high levels of practically every necessary amino acid and mineral, including iodine, phosphorus, potassium, iron, copper, and vitamins A and D. Aquaculture contributes to global food security, and its supply has been steadily expanding in recent decades. Fish supplies underprivileged people in society with nutritionally appropriate food, allowing them to improve their diet quality. Fish farming has been conducted in East Africa for a long time, mostly to meet nutritional needs and, to a lesser extent, to generate revenue. In East Africa, integrated aquaculture is one of the most viable solutions to food shortages and malnutrition.

In terms of nutrients, integrated farming systems are frequently compared to less diversified and more open systems, such as monoculture systems. The diversity of firms in integrated agro-aquaculture is thought to reduce perils. Furthermore, increasing internal recycling and reducing reliance on external inputs are frequently regarded as more sustainable. Integrated agro-aquaculture usually entails on-farm waste recycling or various resource usages to increase production capacity, as well as environmental improvements. Manure loading from animal wastes boosts productivity as a result of livestock integration with fish. The direct application of fertilizer, which increases the generation of plankton in the fish pond, is a common method for enhancing fish productivity in ponds. The utilization of substitutes of crops, livestock, and fishes, as well as the recycling of farm renewable resources and natural resource conservation, is the primary premise involved in integrated agro-aquaculture. Integrating fish farming with irrigation systems increases resource efficiency. Organic manure from poultry, cattle, pigs, goats, sheep, and ducks has been used as the primary source of fertilizer in fish ponds by fish farmers. When organic fertilizers are introduced to ponds in tropical semi-intensive aquaculture, they have a long history of increasing fish production. It promotes the growth of natural food in fish ponds by supplementing the plankton with vital lacking elements. The goal of fish pond fertilization is to improve water quality and increase the variety and amount of phytoplankton and zooplankton, which leads to increased fish yield and financial benefits. Increased fertilizer use efficiency in fish ponds, increased primary productivity, dissolved oxygen, pH, and total phosphates are all positive benefits of fertilization on pond production.

2.5 NUTRIENTS

Nutrients can be obtained from a variety of sources, including air, water, and soil, and each of these sources can give a variety of critical nutrients. Fertilized fish ponds include nitrogen, phosphorus, and potassium, which are used to fertilize agricultural crops. Pond effluents have been used as irrigation water for crops. Plants require some important elements in order to grow normally and complete their life cycle. Carbon, hydrogen, and oxygen are non-mineral elements that are used in the greatest quantities. Plants only take up other elements from the soil in mineral form. N, P, and

K are all essential nutrients for plants. Organic farming has been shown to enhance total nitrogen, nitrate, and accessible phosphorus levels in soil while also preventing nutrients from leaching. The limiting factors in crop growth, development, and ultimately economic output are nitrogen, phosphorus, potassium, and water; nonetheless, plants require the most nitrogen of the three major nutrients. On the other hand, although nitrogen promotes quick growth, enhances leaf size and quality, hastens crop maturity, and promotes fruit and seed development, phosphorus is required for regular plant growth. Such nutrients can be found in the water from a fertilized fish pond. Environmental management and multi-products within a production unit are rationalized by agro-aquaculture. Agro-aquaculture can purify aquaculture effluent by absorbing nutrients that would otherwise build up or be discharged into the environment and cause environmental damage(Turcios & Papenbrock, 2014). This lowers the aquaculture system's wastewater effluent costs as well as environmental damage costs.

Through the integration of resource flows within the system, integrated agro-aquaculture improves on-farm resource usage efficiency and productivity. The abundance of plankton was observed to be higher in ponds receiving chicken dung and it is due to the accessibility of nutrients supplied by poultry manure and supplementary feeds, such as nitrogen and phosphorus. It was found that ponds containing organic manure had higher plankton abundance than ponds without manure. A rise in plankton biomass is frequently linked to nutrient enrichment. Plankton are sensitive to changes in nutrition levels. When comparing vegetable plots irrigated with water from fish ponds to vegetable plots irrigated with water from a stream, the findings revealed that vegetable plots receiving water from fish ponds had greater growing performance and yield. In comparison to plots irrigated with stream water, the length, diameter, and number of leaves of Chinese cabbage, as well as yield, were all higher in plots irrigated with water from fish ponds. This is most likely owing to increased nutrient contributions from the fish pond as a result of inputs like manure and feed. Poultry dung has been reported to be the most effective approach of delivering nitrogen and other critical elements to fish ponds. Ponds treated simply with manure and vegetable plots watered with water from these ponds, on the other hand, had the highest benefit cost ratio (BCR). Fish feeds account for around 60 to 70 percent of fish farm operation costs(Baki & Yucel, 2017). Applying manure to fertilize ponds may reduce the need for extra feed while increasing productivity and profitability.

The Nile tilapia is an African cichlid fish. It has been successfully farmed in a variety of environments and is a significant aquaculture fish species in many areas of the world, especially in tropical and subtropical countries. It is regarded as one of the most productive and widely traded food fishes on the planet. The development of tilapia culture has received a lot of attention in recent years. The global production of farmed tilapia is increasing, with 6,510,700 tonnes produced in 2017("Fisheries and Aquaculture - Cultured Aquatic Species - Oreochromis Niloticus," n.d.). Tilapia is one of the most significant cultivated fish in Nigeria. The fact that *O. niloticus* develops and reproduces in a wide range of environmental conditions, has a low trophic level for feeding, and can survive stress caused by handling and poor water quality all contribute to its success. One of aquaculture's top concerns is to expand fish production to meet the demands of an ever-increasing population. Nigeria, like many other developing countries around the world, is struggling to meet the protein needs of an ever-growing human population. The production of capture fisheries for both food and non-food purposes has plateaued. According to FAO, the world's seas' maximum wild catch fisheries potential has most likely been reached. Because fish is a low-cost source of animal protein with no religious or cultural restrictions on its eating, boosting aquaculture output in Nigeria is plainly needed, especially while capture fisheries resources are depleting due to overfishing, habitat damage, and pollution. Because fish farming is a business, correct management strategies and economics must be implemented. The cultivation of fish in a controlled environment, vats, and plastics is known as fish farming. Aquaculture needs to be expanded and the output of aquaculture needs to be improved. The most common structure for aquaculture production is a pond. *Clarias gariepinus*, (Figure 2.5) *Heterobranchus* sp., and their hybrids are the most commonly farmed species in Africa, particularly in Nigeria. In Nigeria, fish culture in plastic and concrete tanks is becoming common. Fish farmers work in concrete tanks that vary in size and design according to individual preferences, available space, and financial resources.

Figure 2.5: Clarias gariepinus.

Source: https://upload.wikimedia.org/wikipedia/commons/7/75/Clarias_gariepinus.jpg

The majority of this output is destined for local markets, primarily as a low-cost, high-quality animal protein source in rural areas. Little attention was paid to tilapia harvesting and processing because it was considered a low-value product. According to the International Standard Statistical Classification (ISSC), tilapia has not been classified as a commerce item. There has been a shortage of data on tilapia exports and imports. However, the worldwide trade in tilapia and tilapia products began to attract more interest from around the world. In order to adopt the most cost-effective production strategies, it is necessary to shed some light on the aspects that may affect tilapia production, costs, trade, and overall economics, with a focus on harvesting, handling, processing, and economics. Small-scale rural farms dominate tilapia culture, especially in developing nations, and the majority of production is consumed locally. For farmed tilapia, there is no specific harvesting process. There is also a lack of information about tilapia harvesting methods. Furthermore, polyculture with other aquatic species makes adopting specialized tilapia harvesting techniques more challenging. As a result, small-scale farmers in various places may use a variety of simple harvesting techniques.

2.6 HARVESTING

The harvesting methods used are determined by the size of the pond, the culture systems used, and the level of technology used. Small-scale tilapia growers, for example, in many countries of Africa and Asia, find it difficult to get harvesting nets and other equipment necessary for comprehensive pond harvesting. To address this issue, several farmers use partial harvesting techniques, which make use of locally accessible equipment. Because they favor batch harvesting, large-scale tilapia producers use more modern harvesting instruments like winches. To decrease stress and comply with animal welfare rules, they sometimes anesthetize the fish before harvesting. Partial harvesting is frequently done to remove huge fish and make room for lesser fish to flourish. Because of the limited information supplied, partial harvesting may be a more cost-effective harvesting strategy than single harvest procedures. Researchers compared partial and batch harvests of Nile tilapia grown in monoculture. Monoculture and integrated systems with several harvesting procedures produced more total fish than a monoculture system with a single harvest. Despite better yields, the integrated system with multiple harvest yielded a lesser return than the monoculture system, most likely due to the higher labor cost.

When mixed-sex tilapia was treated in various harvesting approaches, it was discovered that the lowest yield was obtained when no partial harvesting was carried out and the fish were taken only once after 365 days(Saiti, Jamu, Chisala, & Kambewa, 2007). On the other hand, when partial harvesting was used every 90 days, commencing with a first harvest of 60 g fish, the highest yield was attained. However, the highest earnings were obtained when the fish were harvested in stages over 120 days, with the first harvest occurring on day 90 and eliminating all fish weighing 60 g or more. The annual output of all-male tilapia grown in polyculture with juvenile freshwater shrimp (Figure 2.6) and partially harvested monthly was similar to that of a batch harvest technique. Partial harvesting has a lot of potential for small farmers that need to supply markets on a regular basis. This approach also makes it easier to meet the demand for fish fry because not all of the fish for stocking are needed at the same time. Again, technological, biological, economic, and sociological issues must be considered while analyzing tilapia harvesting methods. Fish harvesting methods are one of the primary impediments to the growth of rural aquaculture in many rural locations.

Figure 2.6: Freshwater shrimp.

Source: https://www.seafdec.org.ph/wp-content/uploads/2020/03/ED7A4188-495x330.jpg

2.7 MARKET

Tilapia is commonly sold fresh or iced in rural areas and local marketplaces in poor countries, with minimal handling and processing. The global tilapia market, on the other hand, is rapidly expanding, not only in nations like Southeast Asia, the Far East, and Africa, but also in developed regions such as the United States, Europe, and Australia. To maintain quality and meet the quality requirements required by external markets, more attention should be devoted to tilapia handling and processing. Poor handling and holding conditions, insufficient processing, and the adoption of ineffective processing technologies can all have a negative impact on tilapia quality and lead to increased postharvest losses. Insufficient handling and processing resulted in the loss of more than 20% of Nile tilapia and African catfish yields in Nigeria. In many parts of the world, consumers prefer to buy live rather than iced or frozen fish, hence live tilapia has become a popular commodity. In places like Singapore, Thailand, and Malaysia, live tilapia is becoming more prevalent at display tanks and aquaria in seafood restaurants and shops. However, appropriate support systems, such as live holding containers, specially equipped trucks, live holding centres, and other infrastructure components, are required for live food systems.

Transport, distribution, exhibition, and holding facilities were among the phases and processes in the handling and marketing of tilapia in Malaysia. To decrease stress, the fish can be transferred into tanks at a temperature of 21°C for roughly 10 hours(Basri et al., 2020). To limit fish activity, the water in the tanks is gradually chilled to 11°C with the addition of ice. The fish are placed in holding facilities when they arrive and left to resume their normal physiological functions when the temperature rises. This technology allows live tilapia to be transported over a 12-hour travel with no mortality. To reduce stress and increase the quality and shelf life of the fish and their products, several handling, slaughtering, and pre-processing techniques have been recommended. Physical qualities and quality of harvested fish may be influenced by factors such as pre-slaughter circumstances and slaughter weight.

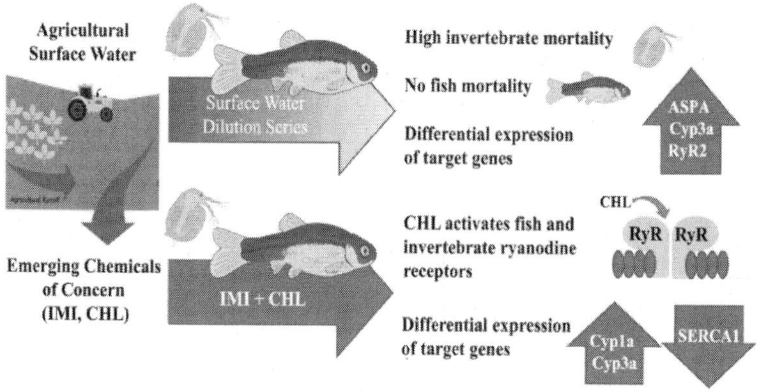

Figure 2.7: Ryanodine.

Source: https://ars.els-cdn.com/content/image/1-s2.0-S0048969721059982-ga1_lrg.jpg

Ryanodine (Figure 2.7) is a substance that permits calcium to be released from the endoplasmic reticulum (ER) lumen via the ER ryanodine receptor. The scientists discovered that transportation time alone lowered cortisol levels for up to 180 minutes, followed by an increase in WHC and RyR1 and RyR3 mRNA expression(Goes et al., 2015). These findings show that pre-slaughter stress can cause an increase in WHC in tilapia filets. Because of the growing demand for processed fish in worldwide markets, fish processing is becoming increasingly significant. Processed freshwater fishes, available in a variety of basic forms, have traditionally enhanced consumer choice. Small processors usually offer their goods directly to consumers. Processed fish, typically in dried and salted form, is marketed along highways in several Thai regions, for example, to accommodate domestic travelers. Large fish processors, on the other hand, typically offer their products to Bangkok's wholesale fish markets as well as shops. Processed fish products like fish balls and minced fish, are occasionally resold in supermarkets. Processed tilapia is also becoming increasingly popular in various Southeast Asian countries. Filet tilapia is exported to the United States from Indonesia and Thailand. These food items are more practical for families with working members who may not have time to clean and cook fish when they get home late from work. Tilapia can be an excellent raw material for producing high-value items. This strategy is likely to result in more jobs, increased income, and a higher standard of living, particularly in rural areas.

2.8 PROCESSING

Tilapia is typically processed using techniques like cooling, fileting, radiation and industrial processing that involves drying, smoking, canning, fish balls and sausages, and so on. In order to produce a high-quality result, a number of precautions must be followed during the production. The following elements often influence tilapia marketing. They are production systems, operating costs, producer competitiveness, degree of processing, production scale, and consumer demand. Because of the high cost of labor, energy, and operating expenditures, manufacturing costs in the United States and Canada are greater than in Central America, Latin America, the Caribbean, and other tropical countries. It would be more profitable to import tilapia from these regions than to farm them in North America. This gives tropical producers a competitive lead over producers in the United States and Canada. This disadvantage can be overcome by improving production management and optimizing resource consumption, such as employing fertilization for the generation of tilapia fingerlings. Tilapia are mostly cultured in semi-intensive systems with low operating costs and few other inputs. Many countries employ intensive tilapia culture with high investment and maximal inputs, albeit on lower scales. As a result, the predicted outcomes of culture practices are a function of the culture systems used and the inputs to these systems. Numerous studies have been conducted to assess and analyze the economic returns of various farming systems in various geographic locations. Such evaluations would undoubtedly aid fish farmers and farm managers in adopting the most environmentally friendly and commercially viable strategies. In brackish-water ponds in the Philippines, researchers evaluated the economic feasibility of the polyculture of tiger shrimp with Nile tilapia.

Polyculture of shrimp and tilapia was lucrative up to a 20% drop in the selling price of both species, according to a sensitivity analysis(Martínez-Porchas, Martínez-Córdova, Porchas-Cornejo, & López-Elías, 2010).

Investment costs accounted for around 5% of total operating expenses. When fingerlings were obtained from backyard ponds or the hapas, fingerling costs that are about 25% of overall expenditures may be greatly lowered(Nasr-Allah, Dickson, Al-Kenawy, Ahmed, & El-Naggar, 2014). Feed expenditures, which accounted for almost 70% of overall costs, may be cut by optimizing the feeding regime and using home-mixed fish feed, as well as extending the rearing time to 100 days. Rice growers might raise their household income by around 20% in this method. Polyculture of Nile tilapia in earthen ponds was found to be more cost-effective than other culture systems when considering local support services such as fry and fingerling supply and local market development. In the United States, a computer simulation model was used to evaluate the economic and engineering aspects of tilapia production in a small recirculating culture system. Sensitivity analysis revealed that of all the operational factors studied, lowering feed costs and increasing FCR were the most effective in lowering production costs. Increased system profitability was achieved by increasing output capacity or lowering investment costs. Feed, processing and distribution, and sex reversed fry were the biggest variable expenses ion a commercial scale. The use of locally prepared feed, as well as the integration of hatchery and grow-out operations, is recommended for lowering operational costs and boosting return. When air charters were used, the most variable cost was freight shipment to Miami, Florida. The approach is economically viable if fry can be sold all year, albeit seasonal market demand may restrict the returns.

The economic and technical feasibility of cage culture in Ghana resulted in the cheapest and most readily available local inputs, the sensitivity analysis revealed that the internal rate of return was 58 percent, indicating that the investment in tilapia cage culture in Ghana is highly profitable, particularly in rural areas(Ofori, Abban, Karikari, & Brummett, 2010). Tilapia farming is primarily done in semi-intensive systems in clay ponds in Egypt, which is a major tilapia producer. At various scales, intensive tilapia farming in ponds, tanks, and cages, as well as recirculating aquaculture systems, is also employed. The cost-benefit ratio of tilapia farms varies greatly and is influenced by the culture system, input availability and pricing, and geographic location. All farming systems, on the other hand, appear to be cost-effective, with variable degrees of profitability. The benefit-to-cost ratio (BCR) in semi-intensive systems ranged from 1.3 to 1.5. BCR is significantly better in intensive culture in ponds and cages(Vineetha Valsalan, Kripa, Padua, Narayanakumar, & Vysakhan, 2020). This is mostly due to intensive tilapia farmers' superior access to cash, technology, and production inputs, allowing them to operate systems with significantly larger operational and capital expenditure. Small-scale semi-intensive farmers, on the other hand, have limited access to funding and production inputs, resulting in low total expenses, income, and BCR; yet, this approach is still quite profitable for these farmers.

These observations clearly show that system analysis is required for successful aquaculture techniques in rural areas. In other words, cultural inputs that are restrictive in one location may not be so in another. As a result, system analysis would present the farmer with various cultural scenarios from which he/she could choose the most cost-effective one.

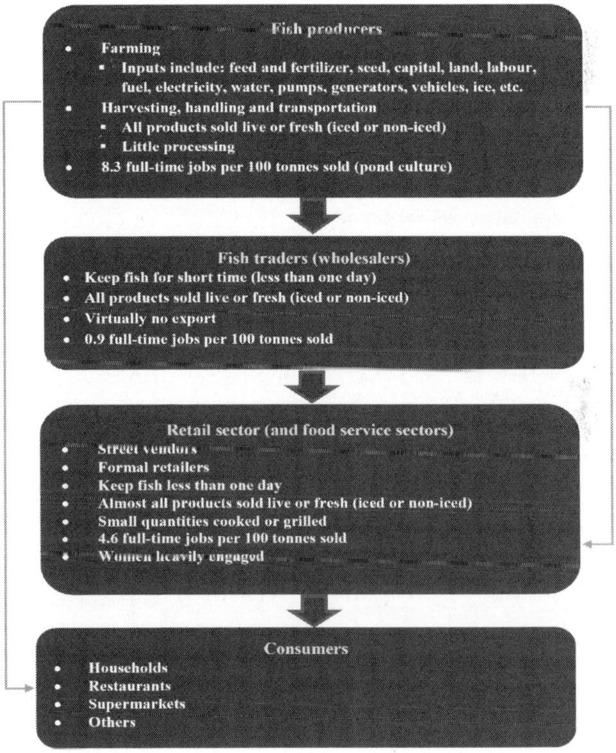

Figure 2.8: Value chain.

Source: https://www.researchgate.net/profile/Emmanuel-Frimpong-8/publication/301220437/figure/fig6/AS:502888485343236@1496909366179/Value-chain-of-farmed-tilapia-in-Egypt_W640.jpg

A value chain (VC) (Figure 2.8) is a series of processes and services that take a product or service from production to end clients, as well as final disposal after usage. All stakeholders, from suppliers to producers, processors, and final end-users, must be included in this approach, which is backed by technical, business, and financial service providers. Value chain analysis (VCA) is a multidimensional analytical technique for comprehending and evaluating the performance of each component of the VC, including product flow analysis, information flow analysis, and overall VC management. VCA has been used to evaluate the performance of various systems, such as factors affecting competitiveness, the costs and earnings of each VC cycle, identifying and analyzing gaps and weaknesses in VC performance, and suggesting appropriate upgrading, management, and development strategies to improve VC performance.

Aquaculture is rapidly increasing and is an essential food production system. It is vital to analyze aquaculture VC, with a focus on the synthesis of new information and viewpoints. VC in aquaculture provides a variety of advantages, including providing policymakers, fisheries managers, and aqua farm managers with a systematic approach to evaluate and assess the industry's processes and added values. In fisheries and aquaculture, value addition is a crucial idea since it generates higher profits, more stable market circumstances, job possibilities, and product variety. Despite tilapia culture's prominence as a significant performer in worldwide aquaculture production, market, and trade, just a few studies have looked at their VCs in a few nations. The tilapia VC, for example, is simple and brief but quite effective in Egypt. From harvest to end consumer, it takes only 1 to 2 days, with minimal post-harvest loss. Because most manufacturing locations are located near densely inhabited areas, the local distribution system is also extremely efficient. Before reaching the final end-user, tilapia VC in Egypt is divided into only three categories: fish producers including input production, farming, harvesting, handling, and transportation, fish traders, and retailers. Except for fish farmers, they discovered that all key chain performers like input suppliers, merchants, and food services enjoy positive profit margins(Anane-Taabeah, Quagrainie, & Amisah, 2016).

The majority of the profit created along the supply chain goes to the input providers. It was stated that establishing well-defined financial transactions with customers, durable contacts with customers, and a robust information management system can improve the efficiency and profitability of tilapia VC in Ghana. To be more lucrative, fish farmers should lower their variable expenses and price their fish using a cost plus or percentage markup. There were identified specific areas of intervention and development opportunities to achieve food security and nutrition demands, as well as safety, quality, and trade benefits. The majority of tilapia supply comes from cage culture, and all farmed tilapia items are consumed locally. The demand for tilapia is rising in the local market, particularly among the middle and lower classes. Farmers sell to wholesalers/retailers, as well as consumers and restaurants in neighboring villages or secondary towns, through the marketing channel. The processing procedures used are only sun drying and small-scale smokers. The export potential exists for processed farmed tilapia, particularly smoked and salted fish. Researchers assessed the VC for tilapia and catfish products in Kenya, as well as women's participation opportunities. Input supply is a critical restriction for all VC participants in the aquaculture business. Males dominate tilapia and catfish production; however, women are fully involved. Harvesting, post-harvest fish handling, and marketing are the only activities in which women are allowed to participate. Some studies found female engagement potential in various sectors of tilapia and catfish VC. Women's participation is determined by their qualities and goals. Female participants, for example, should consider fish selling if they want flexibility and liquidity, while those with high educational and technical backgrounds who want financial strength should consider input supply, and those who want long-term stability can consider fish farming.

2.9 CONSUMPTION

The primary players participating in the VC of farmed Nile tilapia in Tanzania were also identified, categorized, and mapped, as well as the areas in which VC may be upgraded and improved. The marketing margins of various sub sectors along the VC, from tilapia producers to consumers, were also examined, and important limitations influencing various performers were discovered. Water scarcity, unavailability and/or high costs of farming inputs in terms of seeds, feeds and fertilizers, gas, electricity, insufficient knowledge of fish farming techniques, marketing issues, and low security are among the key challenges faced by tilapia growers in the country. Farmers need training in farm-made feed manufacture utilizing locally accessible components. Farmers should be given financial capital help, and performers at each stage of the chain should be grouped in order to access loans(Chenyambuga, Madalla, & Mnembuka, n.d.).

The findings revealed that tilapia demand in local marketplaces may not be met in the short term, implying that the market will swallow all tilapia produced. In the long run, market demand may be saturated, resulting in a reduction in sale prices. The region should look at tilapia export, as this would allow processing plants to participate in the tilapia filet market. To reach the export market, however, a significant amount of work is required, including basic research on export competitiveness, the construction of support industries to use filet by-products, and market promotion. The biggest constraint for tilapia in the local market is that sale prices are significantly higher than those in the international market. Tilapia is raised in polyculture systems with other native fishes, primarily in tiny ponds on home farms, where women are heavily involved. Small farmers rely on their own resources and have limited access to capital, whereas large farmers with sufficient land can obtain bank loans.

CHAPTER 3

EPIDEMIOLOGY OF TILAPIA DISEASES

CONTENTS

3.1 INTRODUCTION

Tilapia has several advantages, including ease of maintenance, rapid development, and the ability to adapt to a variety of environments. Fish life will undoubtedly be stressed as a result of the stocking density. Competition for dissolved oxygen, food competition, and mobility space can all be impacted by high fish stocking density. Fish with a low stocking density will have more room to migrate than those with a high stocking density. The stress level experienced by the fish, which can lead to a decline in the immune system, is one of the effects of fish stocking density. Fish with a weakened immune system are more prone to infections. Parasites are categorized into two types based on the location of the infected organ. They are ectoparasites and endoparasites. Ectoparasite attacks are caused by excessive fish density in aquaculture ponds, a lack of nutrition, and poor water quality. High stocking density might induce fish to collide, making it easy for ectoparasites to spread to other fish. In general, ectoparasite infection in fish does not cause as much harm as infection with other pathogens such as viruses and bacteria, but it is one of the sources of secondary attacks that can result in viral and bacterial infections. Ectoparasites that attack cultured fish can have an impact on their survival, as their growth is slowed. Furthermore, ectoparasite infections in fish result in a reduction in the quality and quantity of products, resulting in financial losses for producers. The types of ectoparasites discovered in tilapia in Hiung Village cultivation ponds are Dactylogyrus, Oodinium, Lernea, Gyrodactylus, and Trichodina. Dactylogyrus had the highest prevalence at 86.67 percent, while Gyrodactylus and Trichodina had the lowest at 6.67 percent(Sumuduni, Munasinghe, Chandrarathna, & Amarasinghe, 2015).

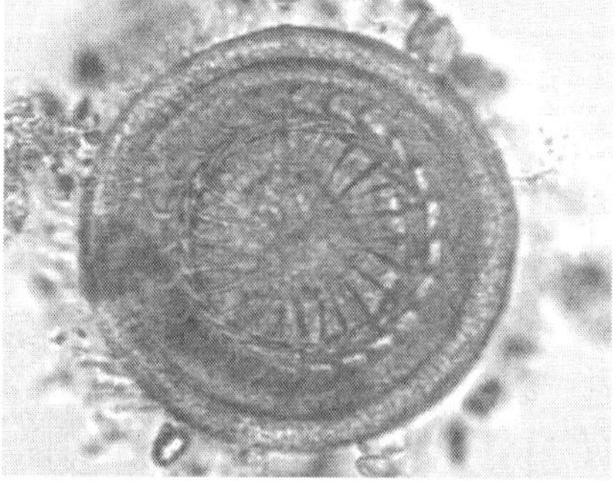

Figure 3.1: Trichodina nobilis.

Source: https://www.nies.go.jp/chiiki1/protoz/picture/inaki/tric-236.gif

3.2 INFECTIONS

Trichodina nobilis (Figure 3.1) is a parasite that is commonly seen in freshwater fish. Because mucus, which is both food and a live ectoparasite, is abundant on the body surface of fish, *Trichodina nobilis* can be found there in large numbers. Furthermore, the body's surface is in direct contact with the environment and is vulnerable to *Trichodina nobilis*. This demonstrates that the fundamental element in the propagation of a parasitic process in the waterways is an interaction between the environment, fish, and parasitic diseases, rather than the stocking density of fish. If the interplay of above and below in layers, which eventually results in cash movements, is supposed to induce stress in fish, the immune system of the fish will be compromised, allowing parasites to readily penetrate the fish's body. Tilapia is the world's third most popular fish, and its high protein content, mild flavor, and ease of preparation are making it more popular in aquaculture. Disease concerns are the leading source of economic losses in aquaculture, particularly in intensive culture mode. Streptococcicosis is a well-known tilapia disease caused primarily by *Streptococcus agalactiae* and *Streptococcus iniae* infections (Soto et al., 2015). Traditional pathogen detection techniques included a variety of methods, the majority of which were time-consuming and labor-intensive. Antibiotics have long been used to prevent and cure infectious disorders in aquaculture, including streptococcicosis (Figure 3.2). The uncontrolled and indiscriminate use of chemotherapeutic medicines, on the other hand, has resulted in the formation of a slew of antibiotic-resistant bacteria. As a result, alternatives to traditional disease detection and control procedures are urgently needed not only for disease control but also for the aquaculture industry's long-term development.

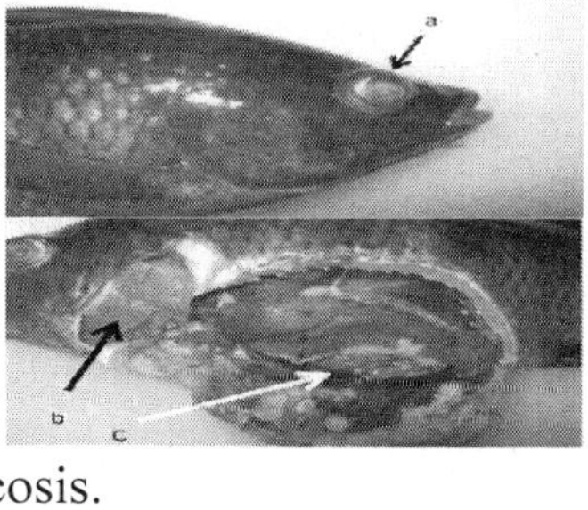

Figure 3.2: Streptococcicosis.

Source: https://www.researchgate.net/profile/Dolores-Baxa/publication/268233632/figure/fig4/AS:271765314207760@1441805308579/Nile-tilapia-Oreochromis-niloticus-from-Lake-Sentani-showing-gross-clinical-signs-of_W640.jpg

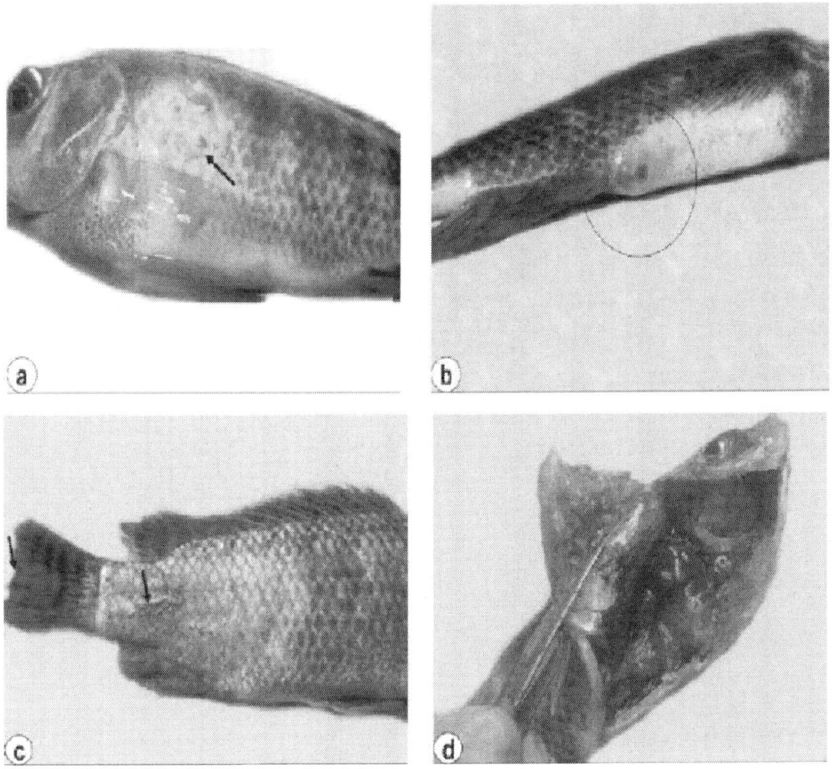

Figure 3.3: Pseudomonas fluorescens.

Source: https://www.researchgate.net/publication/328589864/figure/fig1/AS:687266221527040@1540868444239/Skin-infection-with-Aeromonas-sobria-in-carp-Cyprinus-carpio-L-photo-A_W640.jpg

In South China, tilapia is a popular commercial freshwater fish(Kuwaye et al., 1993). Diseases, on the other hand, are becoming increasingly serious as the culture develops. Bacterial infections such as *Pseudomonas fluorescens* (Figure 3.3), *Edwardsiella tarda* (Figure 3.4), *Aeromonas hydrophila* (Figure 3.5), *Streptococcus iniae*, and *Streptococcus agalactiae* cause a wide range of diseases that cause significant harm to the culture. Antibiotics are effective in the prevention and treatment of tilapia disease, but they have a number of drawbacks, including drug resistance, and pollution.

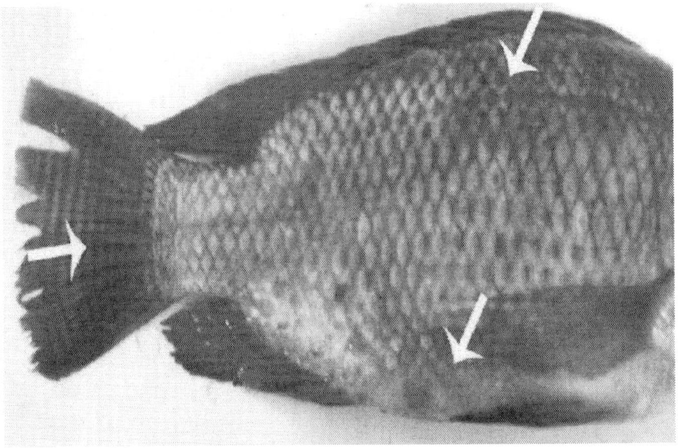

Figure 3.4: Edwardsiella tarda.

Source:https://www.researchgate.net/profile/Haitham-Adam-2/publication/327155923/figure/fig1/AS:662369470525442@1534932596524/Photo-1-O-niloticus-infected-with-E-tarda-showing-body-darkening-hemorrhages-on-the_W640.jpg

Figure 3.5: Aeromonas hydrophila.

Source: https://www.researchgate.net/profile/Mohamed-Fathi-2/publication/340163856/figure/fig4/AS:873020113571842@1585155623480/Experimentally-infected-O-niloticus-with-Aeromonas-hydrophila-complex-showing-grayish_W640.jpg

It is critical to investigate safe and effective disease management measures in order to promote the long-term and healthy development of tilapia farming. We can employ molecular biology methodologies to develop PCR methods for rapid detection of *Streptococcus agalactiae* in order to detect tilapia bacterial infections rapidly and reliably. Because of their lipid membrane, viral particles have been found to be susceptible to organic solvents like ether and chloroform. Although the duration of life outside the host has yet to be defined, horizontal, waterborne spread has been demonstrated under experimental conditions. Experimental infection has resulted in clinical disease that is similar to that seen in natural epidemics, including substantial mortality rates within 10 days of infection. In situ hybridization produced positive signals in numerous organs like liver, kidney, brain, gills, spleen, and muscular connective tissue in samples taken from Thailand, with the indications in the liver, kidney, and gills. A viral preference for the liver and gastrointestinal tract was found in Ecuadorian samples, with an evident tropism towards hepatic epithelium. In Israel, mortality rates of over 80% have been reported in affected farmed populations, but no such high levels of mass mortality have been documented in wild stocks from whom positive samples have been acquired. Mortality rates ranging from 20% to 90% have been reported in Thailand, with peak mortality rates occurring within two weeks after transfer to grow-out cages (Suresh & Lin, 1992).

Figure 3.6: TiLV.

Source: https://www.researchgate.net/publication/324945760/figure/fig1/AS:622545875259394@1525437911638/Clinical-signs-of-representative-TiLV-infected-Nile-tilapia-and-red-tilapia-naturally_W640.jpg

3.3 TILV EPIDEMICS

TiLV epidemics (Figure 3.6) have been linked to huge farm sizes, high stocking densities, and tilapia-mullet polyculture in Egypt. In Ecuador, affected fingerlings were usually identified four to seven days after being transferred to grow-out ponds, with mortality ranging from 10–20 percent to 80 percent depending on the fish strain. From around 20% in farms with mixed stocking of red tilapia and Nile tilapia in earthen ponds in Phetchaburi province to around 90% in farms with Nile tilapia in Pathum Thani province and farms with red tilapia in open floating cages in Chai Nat province, mortality has been reported from Thailand in farms with various species and production combinations (Jansen, Dong, & Mohan, 2019). Co-infections: Flavobacterium, Aeromonas, and Streptococcus were reported as co-infections in TiLV-positive fish from Thailand, as well as external monogenean parasites like Gyrodactylus and Dactylogyrus and ciliated protozoa such as Trichodina. The proportional importance of TiLV and any co-infections has yet to be identified; nonetheless, infection trials have revealed mortality rates of up to 80% in TiLV-infected experimental populations (Dinh-Hung et al., 2021).

TiLV is expected to be a global problem for farmed tilapia, according to various publications from international organizations. Because of the extensive worldwide circulation of tilapia, a broader transmission of the virus is possible. Screening and surveillance initiatives should be aggressively encouraged in several of the tilapia-producing countries. Multiple nations have started official screening and surveillance programs, according to the FAO. FAO also advised biosecurity measures that countries should adopt while translocating live tilapias, especially in countries where TiLV has been detected and in countries where TiLV status is unknown. International participation in such surveillance programs may help speed up the generation of initial knowledge while local diagnostic competence is being built. In nations with major tilapia production where TiLV has not been found, doing a TiLV import risk study should be recommended. Some concerned countries' competent authorities are gathering additional information and conducting laboratory testing to confirm the scientific results that TiLV is present. It's critical to raise awareness about the necessity of reporting, tracking, and mapping any anomalous tilapia mortality. TiLV-related losses could pose a substantial threat to food security in some parts of the world, given the importance of tilapia as a protein source. Furthermore, its presence is expected to have an impact on worldwide tilapia trade.

In order to measure the actual or predicted impact of disease as a result of TiLV infection, socioeconomic impact analyses should be encouraged. Regarding crucial epidemiological aspects of TiLV, there is a paucity of scientific knowledge. In countries where TiLV has been detected, knowledge of viral properties, transmission methods, susceptible host life stages, TiLV survival outside the host, risk factors for disease outbreaks including different production systems, and the presence of any non-tilapine hosts or carrier species must be investigated. The possibility of variance in risk factors between countries or regions should be addressed due to varying factors such as production methods and fish genetics. Given the extensive work that has gone into developing genetically enhanced tilapia lines, susceptibilities to TiLV must be thoroughly evaluated in the field. Similarly, the potential for vertical transmission of TiLV, as well as the possibility of subclinical carrier status, must be determined as soon as possible. To close these gaps in knowledge, descriptive, observational, and experimental investigations should be done.

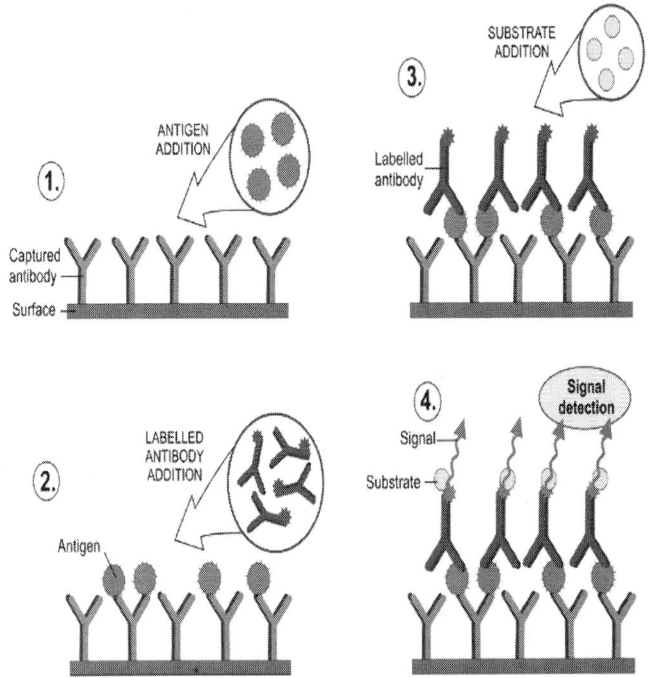

Figure 3.7: ELISA.

Source: https://www.biotek.com/images/applications/ELISA20.jpg

3.4 BIOSAFETY

Improved biosecurity measures, as well as containment programs, should be established based on the collected data in order to lessen the impact of TiLV in the afflicted countries and regions and the risk of further spread to other places. Because TiLV is present on multiple continents, regional capacity building is required across all stakeholder groups.

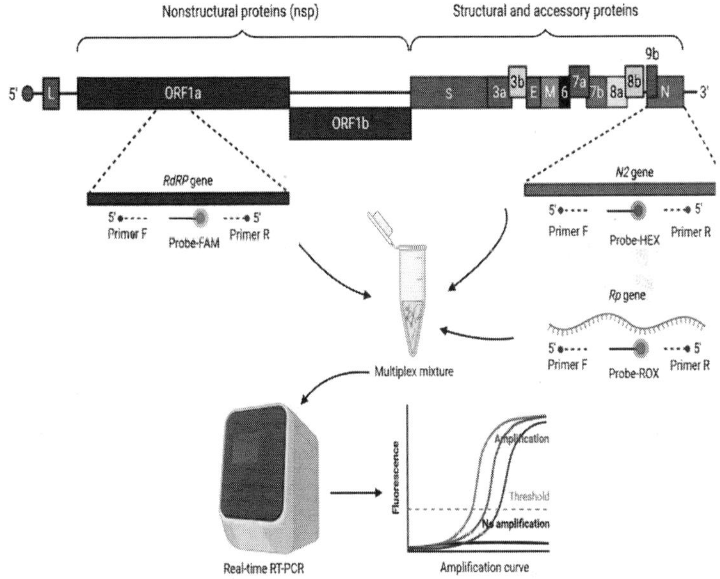

Figure 3.8: Real-time RT-PCR.

Source: https://media.springernature.com/lw685/springer-static/image/art%3A10.1038%2Fs41598-022-06977-z/MediaObjects/41598_2022_6977_Fig1_HTML.png?as=webp

Participatory approaches at all stakeholder levels should be promoted to facilitate the speedy transmission of knowledge (Yamamoto et al., 2014). To improve diagnostic efficiency, strong international collaboration on diagnostic capacity building should be prioritized. To improve screening efficiency and sensitivity, new or improved diagnostic procedures like ELISA, (Figure 3.7) quick antigen strip test, and real-time RT-PCR) (Figure 3.8) should be supported. To reduce the impact of TiLV and the disease it causes, collaborative activities between the corporate sector and appropriate governments should be encouraged. A vaccination that is effective, safe, and economical could be accessible in the near future as a result of continuing vaccine development efforts. Improved biosecurity measures might be combined with an efficient TiLV vaccine and possibly specific pathogen-free fish in the long run. The potential impact of TiLV, together with its widespread dissemination across continents, necessitates strong international collaboration to expand collective knowledge and understanding of TiLV and its implications. Improved TiLV mitigation and control techniques should hopefully be within reach with rapid distribution of new knowledge and effective national and international collaboration. Researchers looked into outbreaks in Israel and discovered a syndrome characterized by lethargy, endophthalmitis, skin erosions, renal congestion, and encephalitis, as well as the disease's transmissibility from affected to naive fish. They cultured a virus from infected fish in E-11 cells i.e., a cloned subculture of the snakehead fish cell line, demonstrated sensitivity to ether and chloroform and used cDNA library screening to find a 420-amino-acid open reading frame (ORF) with no apparent homology to any nucleic acid or protein sequence in existing databases (Bacharach et al., n.d.). The virus was given the name tilapia lake virus (TiLV), but no taxonomic assignment could be made at the time (Figure 3.9).

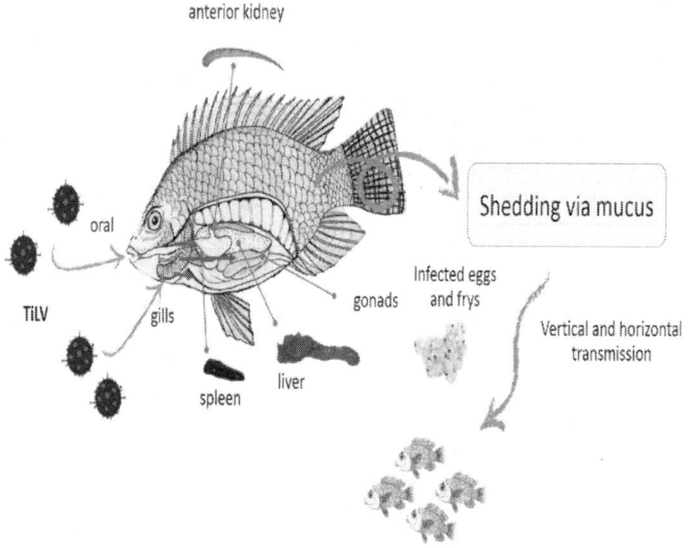

Figure 3.9: Tilapia lake virus.

Source: https://onlinelibrary.wiley.com/cms/asset/5ca7f894-c87b-43f0-bbca-3648a3a204a9/jfd13237-fig-0003-m.jpg

RNA extracts from diseased tilapia brains were depleted of rRNA and utilized as the template for Ion Torrent sequencing in Israel. The template for Illumina sequencing was RNA isolated from nuclease-treated, sucrose gradient-purified, and concentrated particles from infected E-11 culture cells. TaxMaps was used to taxonomically classify reads from two Ion Torrent and two Illumina libraries by mapping against the NCBI nucleotide database, the NCBI RefSeq database, the tilapia reference genome sequence termed Orenil1.1, (Figure 3.10) and corresponding annotated tilapiine mRNA sequences(Conte, Gammerdinger, Bartie, Penman, & Kocher, 2017). The VICUNA assembler was used to assemble unclassified reads. To identify assembled sequences likely to stem from the same segment of the same species of virus in various samples, contigs from each library were matched with BLAST against all contigs from the other three libraries, maintaining hits with an E value of $1e^{10}$ or lower. All of the contigs that exhibited any similarity were grouped together using single-linkage clustering. Ten contig clusters in each of the four libraries were found, with at least one contig in each.

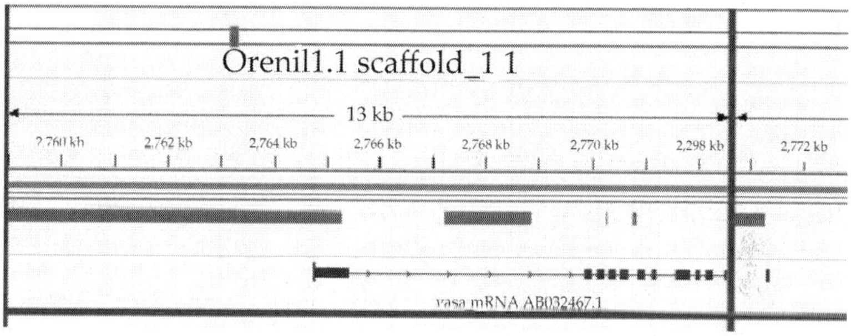

Figure 3.10: Tilapia genome Orenil1.1.

Source: https://www.researchgate.net/publication/345754556/figure/fig1/AS:981696426037252@1611066075834/Vasa-gene-duplication-The-top-row-shows-the-vasa-transcript-NCBI-accession-number_W640.jpg

To account for frameshift issues and infer the longest possible ORF, overlapping projected open reading frames in contigs from various assemblies were employed. We employed a mix of k-mer analysis, read depth analysis, and manual curation to generate 5′- and 3′-terminal sequence motifs to refine terminal sequences, based on a model in which genomic segments are expected to include conserved termini. BWA-MEM mapping of the initial raw read data against the 10 final consensus sequences revealed that 99 percent of unidentified reads from Illumina libraries and 87 percent of unidentified reads from Ion Torrent libraries matched the consensus sequences(Joshi et al., 2018).

To amplify fragments representing all 10 contigs from RNA samples from infected fish and pure virus particles, PCR primers were created and employed. Terminal sequences were recovered in all 10 clusters using 5′ and 3′ rapid amplification of cDNA ends (RACE). We can determine that the clusters comprised 10 viral genomic segments based on similarities in terminal sequences in the various clusters, and will refer to them as segments 1 to 10. With mega blast/BLASTx/tBLASTn or tBLASTx, the other nine regions had no homology to other sequences in the GenBank nucleotide and protein databases. Researchers compared the segment 1 ORF with polymerase subunit PB1 obtained from particular Orthomyxoviridae members based on poor similarity with PB1 motifs. Four motifs conserved in RNA-dependent RNA and RNA-dependent DNA polymerases were found to have low homology in segment 1 putative protein. We can use Northern hybridization, mass spectrometry, and in situ hybridization to discover and correlate the existence of putative viral genes and proteins after traditional sequence homology analysis of sequences obtained from infected fish and cultured cells failed. TiLV is an RNA virus with a genome made up of 10 distinct segments. The PB1 segment of the orthomyxovirus influenza virus C bears the least resemblance to the biggest segment. It also includes the most important polymerase motifs. As a result, we believe that segment 1 encodes TiLV polymerase. The remaining 9 segments have no apparent resemblance to other known viral sequences; however, the existence of complementary sequences at their termini and the identification of proteins in infected cell extracts that correlate with the ORFs they contain give evidence that these are TiLV gene segments. Sequence analysis doesn't reveal anything about their roles in the TiLV life cycle. In situ hybridization tests revealed a transcriptional region in the nucleus.

TiLV is most likely to be classified as a novel genus in the Orthomyxoviridae family. TiLV's 5′ and 3′ non-coding termini include 13 nucleotides that are identical in all segments(Kibenge, 2019). This arrangement is identical to that seen in influenza, Thogoto, and infectious salmon anaemia orthomyxoviruses, and it allows for base pairing and the development of secondary structures necessary for viral RNA replication, transcription, and packing. A short, unbroken uridine stretch can also be seen at the 5′ ends of TiLV genomic RNA segments. This trait is similar to the 5–7 uridine residue stretch reported in other orthomyxoviruses, where the viral polymerase halts when producing poly(A) tails.

TiLV is suspected of being the causative agent of outbreaks of viral encephalitis and syncytial hepatitis in tilapias in Israel and Ecuador, based on the presence of viral nucleic acid at sites of pathology in the brain and liver. The fact that TiLV has been linked to disease in two geographically dissimilar locations shows that TiLV is a global hazard to tilapia aquaculture. We have no proof that TiLV can infect other species; nonetheless, the genetic dissimilarity of TiLV to other viruses suggests that when TiLV sequences are used for phylogenetic analysis, related viruses may be discovered.

3.5 EXPERIMENTS

More confirmatory experiments are needed to isolate the virions. Preliminary RT-PCR results reveal that the virus identified in SHT shares partial genetic homology with TiLV, a new RNA virus that kills tilapia in Israel. Virions were only found in hepatocytes and the Disse space, not in the endothelium, according to ultrastructural analysis. This shows that the SHT virus prefers epithelial cells in the liver over endothelial cells. In contrast to ISA, when the causative isavirus is endotheliotropic, there is no histological evidence of sinusoidal disruption or bleeding. Orthomyxoviruses have been shown to infect a variety of cell types(Dong et al., 2017). Influenza virus infects respiratory epithelial cells, erythrocytes, leukocytes, and platelets, whereas is a virus infects endothelium, as well as leukocytes and gill lamellar epithelium. In contrast to SHT, ISA virus has not been found in necrotic hepatocytes, despite the fact that the is a virus linked to hemorrhagic kidney syndrome infected renal tubular epithelial cells in Canada, implying that ISA virus tropism in epithelial cells is likely in some circumstances. Virion endocytosis was suggested by many cases of cytoplasmic membrane invagination containing virions, with occasional electron-dense coating. Both influenza and isavirus have been demonstrated to require a low pH for the release of their RNA into the cytoplasm, and both have an acidic environment. There were Intraendosomal virions below the plasma membrane in SHT investigation, but no perinuclear, late endosomes. Orthomyxoviruses require nuclear participation during reproduction, which is unusual for RNA viruses. This causes ultrastructural nuclear alterations, such as a rise in nucleolar density, as well as granular material dispersion and presence throughout the nucleus. Nuclear dispersion, granular intranuclear material, and prominence of nuclear pores were seen in sections of tilapia liver infected with SHT virus. These nuclear traits are comparable to those described for orthomyxoviruses, but without more information on the replication cycle of SHT-associated virions, it is impossible to rule out the possibility that they are artifacts(del-Pozo et al., 2017).

The departure from the host cell is the final step in the viral replication cycle. Virus assembly takes place in the cytoplasmic membrane of the host cell in both influenza and isavirus. The enclosed virus particle is then released from the cell, along with the production of lengthy microvilli. Although there was disruption of hepatocyte villi at the sinusoidal pole, instances of viral assembly and delivery were not caught in the sections of this investigation on SHT. To characterize this, more research is required. So far, ultrastructural characteristics of SHT-associated virions have been found to match those of known orthomyxoviruses. The existence of virions within a perinuclear complex of anastomosing membranous structures that substantially resemble the Golgi apparatus is perhaps the most prominent of these. Although the function and importance of this feature in the replication cycle of SHT-associated virus are unknown, it is probable that it represents an intracytoplasmic transit step of virions toward the nucleus or the cytoplasmic membrane. The Golgi apparatus does play a role in the replication cycle of the influenza virus, where it helps transport a subset of viral proteins to the plasma membrane, but this does not involve intraluminal virions. SHT is also distinguished by the production of syncytia, which is not seen in orthomyxoviral infections. In paramyxoviruses, syncytia development is triggered by a fusion factor.

3.6 ISOLATION

The RT-PCR results reveal that the virus identified in TSH and TiLV share a partial genetic similarity. Additional genetic commonality was discovered in a separate investigation. TiLV has been isolated in E-11, a snakehead or *Ophicephalus striatus* cell line, and a primary tilapia brain cell culture. SSN-1 is a cell produced from whole fry, while E-11 is a clone of it, i.e., composed of a mixed population. As a result, it's not unrealistic to believe that E-11 may be used to isolate the SHT virus. This is despite the fact that there may be variances in tissue tropism between TiLV and SHT, as evidenced by gross and histological variety. In TiLV, symptoms include encephalitis and ophthalmitis, as well as minimal hepatic degeneration or hepatocellular pigment accumulation; in SHT, lesions include hepatic necrosis, hepatocyte dissociation or syncytia development, and gastrointestinal mucosal necrosis. Although changes in tissue tropism of one virus are feasible, this variability may indicate that these are separate viruses. In a further investigation, the homology was proven even more. Isolation of the virus, transmission studies, and additional research into its replication cycle and pathophysiology are all questions that need to be answered. In wet-mount preparations, *F. columnare* is a Gram-negative, rod-shaped bacterium that forms columns(Dong, LaFrentz, Pirarat, & Rodkhum, 2015). On a low-nutrient agar media, these bacteria grow in a distinctive rhizoid pattern. Temperature and environmental stress have both been known to cause outbreaks. The majority of columnaris infections show up on the outside as brown to yellowish-brown lesions on the gills, skin, and fins. The lesions may show as a pale area that lacks the usual shine at first. A zone with a pronounced reddish color frequently surrounds these ulcers. Fin erosion is another early indicator of infection. Columnaris disease causes lesions on the back to extend down the sides, giving the impression of a saddle. Lesions on the mouth can appear moldy, and the mouth can become impacted. The filaments breakdown as the bacteria infiltrates the gill lesions, which are often necrotic. Due to a shortage of oxygen, the fish will begin to breathe rapidly and gasp at the surface if their gills are damaged.

Internally, the infection is seen less frequently. During acute outbreaks, germs can enter the bloodstream and cause a systemic infection. Columnaris is a disease that affects fish when they are stressed. Sudden changes in water temperature are likely to cause and hasten the onset of this condition. Poor water quality, poor nutrition, poor handling, and overcrowding are all stressors that can lead to an outbreak. Columnaris can be found in hatcheries, cages, and closed recirculation systems where fry is produced and grow-out facilities. The disease is very contagious once established, and it can move horizontally from fish to fish, resulting in significant mortality rates. Infections can also be spread by the environment, such as through contaminated nets, specimen containers, or food. Columnaris can also cause secondary infections or disorders; for example, columnaris is frequently preceded by winter saprolegniosis. Observation of typical clinical indications such as saddleback lesions or rotting gills and mouth leads to a preliminary diagnosis. Additionally, the bacteria can be seen under light microscopy. The bacteria exhibit a sluggish gliding movement when observed under phase contrast at 400x magnification, forming typical column-like aggregates. Bacterial isolation on a low-nutrient medium and laboratory identification are required for a clear diagnosis.

The best strategy to prevent the spread of columnaris in farmed fish is to reduce stress in the population. Organic wastes feed the bacteria, which can be regulated with regular water changes. The fish will be less stressed if they are fed properly, have acceptable water quality, and are not handled disproportionately. It is critical to disinfect all equipment after each use and to use separate equipment at each rearing facility to prevent the germs from spreading. However, in fact, this can be difficult; however, stress should be avoided as much as possible. To manage the disease in hatchery tanks and limit the risk of infection during shipping, salt might be utilized. In many circumstances, farmers are only able to control outbreaks to a limited extent by employing antibiotics. This, however, is not a long-term strategy. When diseased fish are treated as soon as possible, the best outcomes are attained. Antibiotics given orally to infected fish are often ineffective because they have a reduced appetite. Antibiotics, in general, only prevent an infection from spreading. Once a disease is gone, new infections are less frequent, demonstrating that an immune response can be induced following infection. In the United States, Intervet Inc. has released Aquavac-COLTM, a live-attenuated immersion vaccination against columnaris in channel catfish(Guz, Puk, Walczak, Oniszczuk, & Oniszczuk, 2014). Vaccination against *F. columnare* in tilapia and other species will be accessible as a vital preventative approach in the fight against this terrible disease.

3.7 TILAPIA CULTIVATION

Tilapia cultivation has been linked to a number of diseases. Streptococcosis is the most dangerous disease because it can kill huge fish in great numbers, resulting in significant economic losses(Zhang, 2021). *Streptococcus agalactiae* is the most common cause of streptococcosis in farmed tilapia. *S. iniae* causes death as well, but to a lesser extent. *Streptococcus* spp. are Gram-positive cocci that are non-motile, non-acid fast, oxidase-positive, and catalase-negative. There is no discernible variation in the clinical symptoms caused by each Streptococcus species. Eye diseases, such as endophthalmia or exophthalmia, are common in sick fish. Ocular opacification, either unilateral or bilateral, might be linked to the prior symptoms, and eye hemorrhages are prevalent. Streptococcus infection does not always result in ocular problems in fish. 2 to 3mm abscesses symmetrically positioned on the inferior jaw are prevalent in Streptococcus-infected fish. These abscesses usually burst and turn into hemorrhagic ulcers that do not heal. At the base of the pectoral fins, larger abscesses of about 5 mm can also be seen. Large abscesses are frequently found at the base of the tail. Infected material is present in these abscesses. When fish survive a Streptococcus infection, abscesses often develop and are discovered after processing. External hemorrhages are caused by streptococcosis. Multifocal pin-point hemorrhages are most commonly seen near the mouth or at the base of the fins. A reddish coloration around the anus or on the vaginal papilla might be seen. During acute streptococcosis outbreaks, abdomen fluid is frequently seen. Ascites is frequently seen in conjunction with a bulging anus.

Internal symptoms are similar to those seen in septicaemic infections in general. Dry feed is rarely found in the stomach or gut of infected fish. In pond-cultured fish, however, some diseased fish may still be filter feeding at the start of an outbreak(Abdel-Latif, Dawood, Menanteau-Ledouble, & El-Matbouli, 2020). A large gall bladder is often seen in conjunction with an empty gut and stomach, indicating a lack of digestive action. Bacteria quickly enter the bloodstream and spread to all internal organs during acute diseases. Hemorrhages in the liver, spleen, kidney, etc., and digestive system are major clinical indications of this septicaemic disease. Spleen and kidneys are frequently swollen. Internal organ adhesions to each other and to the peritoneal cavity walls are common in severe infections. The presence of fibrinous material in the peritoneal cavity can also be seen. Outbreaks frequently occur when fish have been exposed to stress for an extended length of time, such as a rise in water temperature, low oxygen levels in the water, or overcrowding. The disease spreads horizontally from fish to fish through cannibalism, skin damage, and other means, as well as from the environment to the fish. Streptococcosis has the potential to infect fish of all sizes. Larger fish between 100 g and market size are more susceptible to the diseases. During the high-water temperature season, the disease can be severe, with mortality peaks lasting 2-3 weeks.

Gram stains can be used to stain internal organ impression smears of the brain, liver, spleen, and kidney. In real cases of streptococcosis, their inspection under light microscopy at 1000x resolution will indicate the presence of Gram-positive cocci(Jantrakajorn, Maisak, & Wongtavatchai, 2014). The identification of the Streptococcus species that caused the death necessitates bacterial isolation and identification in the lab. During acute Streptococcus epidemics, lowering the feed, either partially or completely, can assist control or reduce the mortality rate. Bacteria are present in the water and feeding aids their absorption. In tilapia farming, high production is accomplished by balancing a high stocking density with a high survival rate. When mortality rises, reducing stocking density reduces both stress and pathogen burden in the population. It is recommended that paddle wheels be used to aerate the water in order to avoid low oxygen levels, which will aggravate the condition. High water temperature is both a source of stress for the fish and a favorable environment for germs to grow. As a result, in recirculation systems with temperature control, reducing the water temperature can be an option. Paddle wheels can also be activated at night when the air temperature is cooler to help reduce the water temperature. Antibiotics are only useful in treating a Streptococcus outbreak if they are administered early in the disease's progression. Furthermore, many farmers claim that antibiotics can only partially decrease mortality rates during the application period. However, once the antibiotic course is over, mortality tends to rise again. This effect also leads to unsustainable behavior: as mortality rises following a standard antibiotic course, farmers are inclined to extend the time of application or use larger doses than indicated.

3.8 NEGATIVE EFFECTS

The negative effects of employing antibiotics, including antibiotic resistance and residue problems, must be carefully addressed, especially for export-oriented companies. There are currently no commercially available vaccines that are useful in preventing Streptococcus epidemics in tilapia. Vaccination, on the other hand, will be the primary preventive strategy for combating this disease and reducing its terrible economic impact on tilapia farming in the near future. Among the few viruses that affect cichlids, the Tilapia Lake Virus is the most dangerous. Although *Oreochromis niloticus* is more resistant to the SVCV than susceptible hosts, due to a variety of stress conditions that cause immunosuppression, *O. niloticus* can become infected with the virus and serve as a virus carrier with little or no clinical signs or post mortem lesions. Viruses have lately been linked to a number of significant disease outbreaks involving both farmed and wild tilapia. Infection with betanodavirus and herpes-like viruses caused changes in the central nervous system, and a novel enveloped RNA-virus caused ocular, cutaneous, and meningeal disease in these epidemics. Tilapia lake virus is a discovered segmented negative-sense RNA virus that has been linked to widespread tilapia deaths in Israel and Ecuador. The virus appears to cause brain problems in Israel while attacking the fish liver in Ecuador and Colombia. This virus appears to represent a serious threat to tilapia aquaculture around the world. The syncytial hepatitis of tilapia, which has been described in Ecuador, is a unique cause of tilapia mortality suspected of having a viral etiology. Ascites was present, and histologically, hepatic lipoprotein buildup, necrosis, and syncytia development with gastrointestinal mucosa necrosis were present. Researchers observed that the virus linked to SHT has an ultrastructural resemblance to an orthomyxovirus and shares some genetic homology with TiLV. They concluded that the causative agent is a herpes-like virus based on the pathogen's morphological, molecular, and biophysical characteristics.

Figure 3.11: Viral nervous necrosis disease.

Source: https://www.researchgate.net/profile/Jalil Zorriehzahra/post/Which-condition-should-be-considered-in-Viral-Nervous-Necrosis-VNN-challenge/attachment/5a02a482b53d2f10b0ba79a7/AS%3A558308914614272%401510122626717/image/VNN2.jpg

The viral nervous necrosis disease (Figure 3.11) is caused by betanodavirus, which has been found in numerous cultured marine fish species around the world, as well as freshwater species recently, producing high mortalities, particularly in larval and juvenile fish stages(Bigarré et al., 2009). The VNN infection of tilapia was discovered using an indirect fluorescent antibody test in the striped snakehead cell line, followed by PCR product sequencing. Bacteria are key infections for both cultured and wild fish, and they cause enormous losses in fish production, which is a major economic constraint in intensive aquaculture. Motile aeromonas septicemia is a septicaemic disease that affects warm-water fish in freshwater habitats and can be acute or chronic. *Aeromonas hydrophila*, *Aeromonas sobria*, and *Aeromonas cavieae* were placed inside the motile aeromonas and named aeromonas complex. Smaller fish have more acute infections, whereas older fish have chronic infections. *Aeromonas hydrophila* is the most common cause of motile aeromonas septicemia and the primary agent of cultured tilapia hemorrhagic septicemia. Overcrowding, inadequate nutrition, temperature shock, low dissolved oxygen content, inappropriate handling, and elevated ammonia and nitrite concentrations are commonly related to motile aeromonas infections in fish populations. MAS causes a condition known as post-stocking syndrome, in which fish begin to die 3 days after being transported and stocked into a new body of water.

Researchers identified A. hydrophila from Nile tilapia, O. niloticus, after massive mortalities reached 98 percent at Barsiq farm in northern Egypt after exposure to extremely cold water (Ruangpan, Kitao, & Yoshida, 1986). Pseudomonas fluorescens and Pseudomonas putida are pathogens that cause bacterial septicemia in freshwater fish, particularly in culture settings. Pseudomonads are Gram-negative opportunistic pathogens that are found in the aquatic environment and are part of the normal gut flora of healthy fish. They are difficult to identify from aeromonas-induced septicemia in fish stocks exposed to various conditions. Pseudomonas fluorescens is commonly found in aquacultures and is one of the most common causes of bacterial hemorrhagic septicemia in fish, which is linked to stress factors such as overcrowding, low temperatures, and injuries, as well as a secondary invader of injured fish tissue and chronic virus infection. When the water temperature was between 15-20°C and 43°C, the natural mortality rate was highest. Pseudomonas spp. were discovered in Egypt from a cultured O. niloticus outbreak in the Qaroun and El Rayan lakes (Abdel-Khalek, Zayed, Elsayad, & Zaghloul, 2020). The organisms were detected in 30.83 percent of the fish analyzed, and some of the fish, particularly during the large mortality episode, showed characteristic symptoms of pseudomonas septicemia. Gram-negative motile rods, Vibrio spp., cause fish hemorrhagic septicemia. Vibrio anguillarum and Vibrio vulnificus were the two most common kinds. Vibrio infections are frequent in marine and estuarine fish, and they have also been reported in freshwater fish, causing major mortality in aquaculture. When fish are closed in heavily populated, commercial systems, it can spread quickly, with morbidity reaching 100%. Vibrio vulnificus is a pathogenic bacterium species found in brackish waters in tropical and warm climates (Chen, Wooster, & Bowser, 2004).

CHAPTER 4

GENETICALLY IMPROVED TILAPIA FARMING

CONTENTS

4.1 INTRODUCTION

Tilapia farming has centered on one species, the Nile tilapia. In the 1970s, Asia's tilapia farming prospered, supported by regional breakthroughs in hatchery technique and pond management. Unlike terrestrial animal farming, where selective breeding has been practiced for generations, most fish farming, particularly tilapia farming, has been overlooked until the mid-1980s(Yáñez, Joshi, & Yoshida, 2020). The repercussions of this lack of attention to genetics were starting to manifest in stagnant tilapia yields by that time. The production of GIFT (Figure 4.1) has demonstrated that selective breeding is a viable and cost-effective method for improving the genetics of tropical farmed fish.

Figure 4.1: GIFT.

Source: https://www.globalseafood.org/wp-content/uploads/2020/04/BARRIA-tilapia-Pic-0_1500-1440x1080.jpg

GIFT was produced without using contentious biotechnology or genetic alteration. In the Asia-Pacific area, GIFT has become a key contribution to national tilapia breeding programs and farmed tilapia output. Research methods and protocols that were proved during the development, dissemination, and evaluation of GIFT are now being applied to other commonly farmed species, particularly Asian carps, across Asia and the Pacific and other developing regions. Some studies investigated the initiative's catalytic effects, depicted its varied outcomes and implications, taught relevant lessons, concerns, and possibilities, and made recommendations for further research, development, and dissemination to have a bigger impact on poverty reduction.

4.2 AQUACULTURE

Aquaculture has grown at a faster rate than any other animal food-producing industry on the planet. Aquaculture has provided the majority of the world's recent profits in per capita food fish supplies. Fish is a source of animal protein, healthful fats, and necessary micronutrients for over 1 billion people around the world. Unlike crops and cattle, most farmed fish have just recently been domesticated and genetically improved, and many still closely resemble their wild counterparts. Only in 1982 did the world's first International Symposium on Genetics in Aquaculture take place. Most aquaculture research and development before the mid-1980s was focused on seed production technologies and enhanced fish management rather than genetic enhancement. In the 1970s, Nile tilapia farming flourished in Southeast Asia, particularly in the Philippines and Thailand (Eknath & Hulata, 2009). There were recent advancements in tilapia hatching technologies as well as pond and cage husbandry across the region. These advancements helped to enhance farmed tilapia productivity, masking a lack of attention to their genetics in the 1970s and 1980s. Some of the repercussions of the lack of attention to tilapia genetics began to surface in the 1980s. Between the 1960s and the 1980s, there were roughly 40 consecutive generations of wasted opportunities for genetic development (Baroiller, D'Cotta, Bezault, Wessels, & Hoerstgen-Schwark, 2009).

4.3 GIFT

The purpose of this impact evaluation study or IES is to evaluate the effects of the production of genetically enhanced farmed tilapia or GIFT and their distribution in a number of nations. From 1988 through 1997, the Asian Development Bank or ADB provided technical assistance or TA to promote the R&D and diffusion of GIFT (Madan M. Dey & Gupta, 2000). This would allow for quick findings from breeding studies and the rapid spread of improved breeds. The goal can be to create GIFT utilizing traditional breeding procedures rather than genetic alteration through gene transfer. GIFT was produced with five immediate goals in mind: to develop improved Nile tilapia breeds, build the resources for national institutions in aquaculture genetics research, disseminate GIFT, conduct genetic, socioeconomic, and environmental evaluations of GIFT, and facilitate the development of national tilapia breeding programs. These immediate goals were set in the context of long-term national research capacity building in order to maintain tilapia genetic development and undertake national fish breeding programs. Improved tilapia varieties were projected to boost profitability for fish producers while also increasing animal protein available for rural and urban customers, particularly the poor.

The distribution of GIFT, GIFT methodology, and knowledge on better tilapia farming practices was hastened by networks among countries at various levels (Fan et al., 2017). Diverse dissemination routes have complemented these networks, helping to deepen the relationship between genetics research and the distribution of improved tilapia breeds to farmers. Fish seed traders have also acted as disseminators, assisting in the connection of tilapia seed producers with clients. The development of GIFT and tilapia farming practices has also been aided by farmer-to-farmer relationships and social networks among members of rural communities. To varying degrees, the development and spread of GIFT have aided the expansion of public-private partnerships. Seed production, seed distribution, extension, farm financing, and setting tilapia sector directions are some of the key areas of partnership that have emerged. In the Asia-Pacific area, the development and distribution of GIFT and GIFT-related approaches have significantly increased public investment in the application of genetics to aquaculture in general, and tilapia farming in particular. In the Philippines, Thailand, and Vietnam, tilapia farming has become one of the key foundations for boosting national fish supplies. Fish genetic improvement and breeding initiatives are given a high priority in related policy and planning papers. Following the spread and application of GIFT, national tilapia plans call for significant increases in farmed tilapia production. Such plans have taken into account both rising domestic tilapia demand and small prospective exports.

4.4 GROWTH

The growing prevalence of GIFT and GIFT-derived strains in tilapia seed supply demonstrates their significant impact on farmed tilapia output. In 2003, GIFT and GIFT-derived strains accounted for 68 percent of total tilapia seed produced in the Philippines, 46 percent in Thailand, and an estimated 17 percent in Vietnam, according to a survey (Ponzoni, Hamzah, Tan, & Kamaruzzaman, 2005). Because of a GIFT-based national tilapia breeding program, the overall contribution of GIFT and GIFT-derived strains in Vietnam is predicted to expand significantly. GIFT has yet to make a significant contribution to Bangladesh's national freshwater aquaculture production, but this is likely to change as farmed tilapia become more widely available and popular. GIFT has been extensively used in research to improve the performance attributes of farmed tilapia in the People's Republic of China. The introduction and spread of GIFT have resulted in a huge increase in rural income and employment. Hatchery owners and fish farmers alike find tilapia farming to be a lucrative business. For example, tilapia farming directly and indirectly employs at least 280,000 people in the Philippines and 200,000 people in Thailand, including their families, on a yearly basis (Acosta & Gupta, 2010). Those who gain from employment in tilapia farming and related activities include the impoverished and small-scale farmers. Fish is especially significant in the diets of people in the world's poorest countries, accounting for more than half of their animal protein. One of the Millennium Development Goals, eradicating hunger and reducing the prevalence of malnutrition, requires special attention to fish production and consumption. Tilapia was a good source of nourishment for fish eaters. Tilapia has proven to be a more cost-effective source of protein than pork, chicken, and other meats. The growing usage of GIFT suggests tilapia's current and future contributions to human nutrition, including rural and urban poor in nations where tilapia is farmed.

The production and transmission of GIFT have demonstrated that selective breeding is a viable and cost-effective method for improving the genetics of tropical farmed fish. GIFT was created without using biotechnology or genetic alteration. The GIFT experience has revealed that prior to commercial production, systematic analyses of the performance of genetically modified farmed fish under a variety of situations are required. Assessing economic viability, social acceptability, and environmental compatibility requires multidisciplinary skills. Multilevel collaborations and broad-based networks with common goals and reciprocal commitments are extremely useful tools for developing and spreading genetically modified farmed fish. Global collaborations, as well as national research and seed and broodstock distribution networks, have aided in the widespread adoption of GIFT and GIFT methodologies, as well as the propagation of better hatchery and farming practices. By providing technical assistance, the ADB has aided pioneering initiatives in the development and diffusion of GIFT. The ADB's Technical Assistance has aided in the formation of larger multinational partnerships and networks, as well as the mobilization of additional resources to promote and recognize the importance of farmed fish genetic improvement and national fish breeding programs in the development of aquaculture. In order for the development and distribution of genetically enhanced farmed fish to flourish and be sustainable, certain enabling factors must be in place in a country. These include fish genetics research capabilities, national fish breeding program resources and commitments, fish seed production and distribution networks and partnerships, market-driven demand and attractive returns from fish farming, supportive policies, facilities, and infrastructure for fish farming; and biosafety and environmental safeguards.

4.5 GENETIC IMPROVEMENT

Unlike crops and cattle, most farmed fish have just recently been domesticated and genetically improved, and many still closely resemble their wild counterparts. More than 80% of the world's farmed fish comes from Asia. Until recently, the majority of fish were raised from wild fish seed through fry and fingerlings or the offspring of captive spawners called broodstock with little or no genetic manipulation. The ability to develop consecutive generations of broodstock to sexual maturity and the production of fish seed in hatcheries began in the 1970s for most Chinese and Indian carps, and in the 1980s and 1990s for farmed shrimp and milkfish, respectively. There are no native tilapia species in Asia and the Pacific, and no native fish species with equivalent aquaculture characteristics to tilapia. Small populations of Mozambique tilapia or Oreochromis mossambicus, brought from East Africa to Indonesia as aquarium fish, were the first tilapia transferred from Africa to Asia. O. mossambicus was originally employed in freshwater and brackish water fishponds in Java in 1939. O. mossambicus was widely propagated throughout Asia and the Pacific in the late 1940s and 1950s. The founding stock transferred from Thailand to the Philippines in 1949, for example, consisted of only three males and one female (Paller, Corpuz, & Bandal, 2017). Because of its abundance in Asia, O. mossambicus was a natural choice for Asia's first farmed tilapia. Because O. mossambicus grows slowly and matures at a tiny size, it overcrowds fishponds with small fish. Furthermore, O. mossambicus established itself as an alien, invasive, and occasionally harmful species in natural waters and fishponds used to produce shrimp and milkfish throughout Asia and the Pacific. As a result, tilapia farming in Asia stalled until the Nile tilapia or O. niloticus was introduced from Africa in the 1960s, a faster-growing and easier-to-manage species than O. mossambicus. Most farmed populations of Nile tilapia in

Asia during the 1960s were descended from a single transfer from Egypt to Japan in 1962. Following that, there was sporadic importation of farmed Nile tilapia strains from Ghana and Uganda from Israel, mostly to Southeast Asia, until the early 1980s (Arthur et al., 2010).

In the 1970s, Nile tilapia farming flourished in Southeast Asia, particularly in the Philippines and Thailand. There were recent advancements in tilapia hatching technologies as well as pond and cage husbandry across the region. Because male tilapia grow quicker than females, technology was created to produce 100% male, sex-reversed tilapia seed, avoiding unwanted reproduction, overcrowding, and collection of undersized fish(Costa-Pierce, 2003). These advancements increased farmed tilapia productivity while masking a lack of attention to their genetics in the 1970s and 1980s. Most aquaculture researchers identified their tilapia populations only by species name until the late 1980s, with little or no data about their breeding history. Survival is the most crucial performance feature in tilapia farming, as it is in most aquaculture. A farmer must be able to harvest as many of the fish purchased as seed as possible. Survival to harvest, on the other hand, is dependent on a variety of genetic and environmental factors, as well as interactions between them. Many farmers have overstocked their ponds and cages throughout the history of tropical aquaculture as a precaution against unpredictably high fish mortality. Overstocking has frequently contributed to such deaths, resulting in wasted seed and feed expenditures as well as revenue losses due to lower yields. It is tough to study the genetic factors that influence fish survival on farms. Fortunately, most tilapia are resilient fish that can survive in a variety of farm conditions within their native temperature and salinity tolerance ranges. Because of these characteristics, selective breeding to improve tilapia survival is difficult. However, the developers of GIFT needed to know whether selective breeding could have negative repercussions on survival. Rather than attempting to select for survival, GIFT researchers monitored the survival of several tilapia strains under well-defined circumstances on-station and on-farm.

4.6 DEVELOPMENT

Fast development from seed to harvest size is the most crucial performance feature for most tilapia growers, combined with a high survival rate. The efficiency with which a fish converts feed to body mass is also a significant factor to consider. When the development of GIFT began, however, the overwhelming problem for the majority of tilapia farming systems in Asia was growing fish swiftly to harvest size, even with ad libitum feeding. These farming techniques mostly consisted of fishponds fertilized to create natural plankton as feed and supplemented with nutritious feed, as well as cages where fish were usually fed more formulated feeds. In these circumstances, there is occasionally a temptation to overfeed fish in the hopes of increasing harvests. This makes estimating feed conversion efficiency more difficult. When GIFT was created, there were no methods for genetically improving feed conversion efficiency in farmed fish, and they are currently being developed. Selective breeding has easily increased the growth of livestock and several fish species. Growth rate, on the other hand, is a difficult feature to quantify because fish's growth rates alter as they mature from juveniles to adults. Harvest weight at 90 or 120 days was chosen as the attribute to be improved in the development of GIFT. Given that previous harvest sizes for farmed tilapia in Asia ranged from around 100 grams to about 350 grams, from cropping cycles of up to 9 months, this was a reasonable option as a measure of overall growth rate. These were also convenient times for experimentation: long enough to show variations in growth between tilapia strains yet short enough to allow for repeated trials for selection progress (Trọng, Mulder, van Arendonk, & Komen, 2013).

Some tilapia farmers have observed delayed maturity in Nile tilapia production stocks as a benefit since it would allow them to raise both male and female fish to higher proportions at harvest before they matured and spawned, rather than just all male SRT. Tilapia seed producers, on the other hand, would not generally welcome delayed maturation because it would delay the spawning of their broodstock. Some seed producers could be willing to accept later maturation if the result was female fish that produced larger, more viable eggs and fry that could be sold at higher prices, even if a fish generating larger eggs would ordinarily produce less eggs. Age and size at maturity are challenging qualities to breed in tilapia because of these factors. They are influenced by a variety of genetic and environmental factors, as well as interactions between them. The goal of the GIFT project was to create a general-purpose, faster-growing Nile tilapia strain suited for a wide range of farm conditions, with enough genetic variability to allow for later selection for additional qualities. As a result, harvest weight at 90 or 120 days was the sole attribute that was consistently targeted for improvement. There were a few attempts to select for later maturation, but for the most part, maturation and spawning, as well as survival, were merely observed.

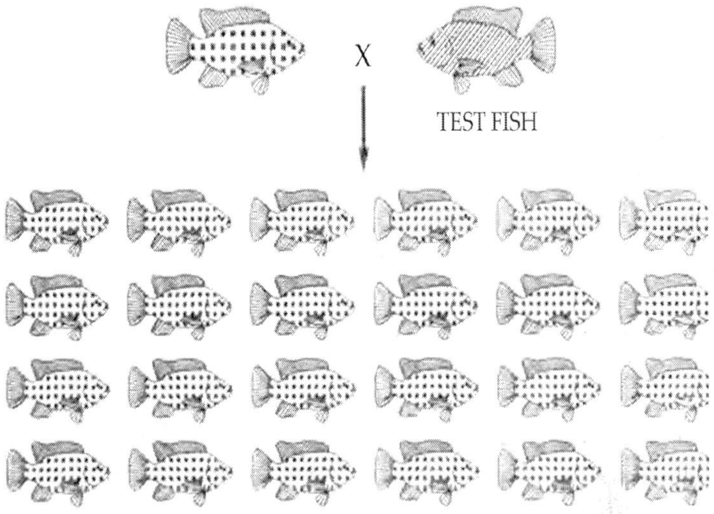

Figure 4.2: Selective breeding.

Source: https://www.researchgate.net/profile/Douglas-Tave/publication/305680808/figure/fig1/AS:421876360060928@1477594570126/Schematic-diagram-of-the-progeny-testing-programme-that-must-be-used-to-create-a_W640.jpg

4.7 SELECTIVE BREEDING

Selective breeding (Figure 4.2) of farmed plants and animals concentrates on moderately or highly heritable commercially desirable qualities. In successive generations, the finest performances are picked as breeders (Rezk et al., 2009). Crossbreeding, also known as hybridization, is a popular alternative to selective breeding because it takes advantage of the unpredictable but sometimes significantly better performance of hybrid children relative to their parents. Hybrid vigor refers to crossbreeds' increased performance. Crossbreeding, unlike selective breeding, which produces gradual genetic enhancements with each succeeding generation, produces a one-time improvement that must be regenerated every time seed is mass produced. This necessitates the upkeep of two sets of parental broodstock. ICLARM and its collaborators predicted that selective breeding would be the most suited technique for the development of GIFT based on the history of genetic improvement in aquaculture, particularly that for Atlantic salmon in Norway (Kuo & Neal, 1982). However, it was also required to investigate the potential benefits of crossbreeding among the eight Nile tilapia strains that had been created.

GIFT should reach harvest size under half the time it takes unimproved tilapia strains in similar conditions. However, further analyses of the performance of GIFT and other Nile tilapia produced in Asia demonstrate substantially smaller advantages for GIFT. This is because the results of any such strain comparison are only valid in the context like place, period, and farming system in which they were collected. Given the region's history of tilapia movements and varying attention to broodstock management, as well as local adaptation of farmed tilapia strains, a wide range of variances in performance between GIFT and non-GIFT tilapia is to be expected throughout Asia. Despite these complexities, the GIFT researchers demonstrated that selective breeding resulted in progressively faster-growing generations of Nile tilapia, and that applying genetics to tilapia farming and, by extension, tropical aquaculture in general can result in significant and rapid development of improved Nile tilapia strains.

The majority of the world's plant and animal breeding and related research is based on large collections of crop varieties and carefully kept livestock breeding nuclei and cryopreserved sperm and embryos, usually referred to as gene banks. Fish gene banks, on the other hand, are uncommon and underfunded, particularly in tropical developing countries. The Nile tilapia broodstock used to generate GIFT, as well as the GIFT synthetic base population and succeeding generations of selectively grown GIFT, make up one of the world's most valuable tilapia gene banks, which is held at NFFTC in the Philippines. The descendants of these fish are still available for national, regional, and international research and breeding through this gene bank. The Philippines' national tilapia breeding program was fully dependent on consecutive generations of selectively bred GIFT until 1999(M. M. Dey & Eknath, 1997). NFFTC then undertook more research and development, deciding to create a new synthetic base population by crossing the current GIFT generation with a non-GIFT strain grown at FAC as well as the original Egypt and Kenya wild strains gathered for the development of GIFT. If these wild Nile tilapia strains had not been available from the gene bank, this would not have been achievable without more costly importation from Africa. The outcome was a GIFT-derived strain known as GET EXCEL by NFFTC, which is currently the foundation of the Philippine national tilapia breeding program, with GIFT methods still being used for selection(Hamilton et al.,

2020). With the help of the Institute of Aquaculture at the University of Stirling in the United Kingdom, a gene bank of cryopreserved sperm was established at NFFTC during the development of GIFT. This is in addition to the gene bank's live broodstock collection. The Philippine government renovated its gene bank facility in 2003 by constructing a new structure. The production and transmission of GIFT have proven to be worthwhile investments with good financial rewards.

Fish seed traders, in addition to research institutes and hatcheries, functioned as GIFT's distribution channels and helped connect tilapia seed suppliers with clients. This was obvious in Thailand and, to a lesser extent, Vietnam, where efficient infrastructure allowed dealers to quickly deliver tilapia seed to farms. In the Philippines, well-functioning markets and distribution networks also aided in meeting the needs of fish producers. The development of GIFT and tilapia farming practices was aided by farmer-to-farmer relationships and social networks among members of rural communities. On a global scale, the foundation of INGA in 1993 aided the widespread diffusion of GIFT, GIFT methodologies, and information(Tayamen et al., 2002). INGA has established worldwide ties, first through DEGITA, which was regarded as the first regional collaborative effort linked with INGA by its national program partners, and then through members sharing their R&D experience and tilapia germplasm. The diversity of INGA's interregional membership, as well as its aid to member countries in sharing GIFT for research, use in national tilapia breeding programs, and distribution to farmers, contributed to increased interest in GIFT and broadened distribution options to many countries.

In 1999, GFII and Genomar ASA of Norway signed an agreement under which GFII and Genomar ASA committed to collaborate on further genetic development of GIFT in the Philippines, with Genomar ASA receiving exclusive rights to commercialize the resulting additional GIFT generations(Hamilton et al., 2020). In exchange for an equity interest in Genomar ASA, GFII transferred tilapia breeding stocks and other assets to Genomar ASA. This business partnership sought to increase the commercialization and distribution of high-performance tilapia seed to fish farmers to match the growing demand for food fish. GFII also kept its own historical GIFT broodstock in order to maintain independent research and development. Genomar Supreme Tilapia, a recognized trademark, has been produced at Genomar-accredited hatcheries in the Philippines from new GIFT generations raised by GFII and commercialized by Genomar ASA.

Genomar's GIFT-based fish seed production and distribution operations have spread to Bangladesh, the People's Republic of China, and Thailand.

To varying degrees, the development and spread of GIFT have aided the expansion of public-private partnerships. Seed production, seed distribution, extension, farm financing, and setting tilapia sector directions are some of the key areas of partnership that have emerged. Public-private partnerships in the Philippines, for example, have evolved beyond tilapia production and distribution to include stakeholders in joint tilapia sector planning. The government and the commercial sector have been working to improve tilapia broodstock quality and maintain connections with a network of hatcheries in tilapia-producing regions in order to ensure a reliable supply of tilapia seed to grow-out producers. The NFFTC of BFAR, its regional and provincial stations, and their hatchery operators and farmer clients have the longest-standing collaborations. However, observers and stakeholders have expressed concern about the prolonged presence of BFAR-operated tilapia hatcheries in locations where private operators dominate because of potential conflicts with private sector commercial interests. In the distribution of tilapia seed, there is now a desire for a better separation of roles between these sectors.

Thailand's government, particularly through the Aquatic Animal Genetics Research and Development Institute, has played a vital role in tilapia breeding research and maintaining high-quality broodstock. Regional genetics centers, which distribute fish to become broodstock at private and public hatcheries, as well as links with the Department of Fisheries' inland fishery stations, which distribute tilapia seed and provide technical advice to farmers, have made public access to GIFT possible in Thailand. Producers are connected to buyers all over Thailand through a well-established network of local and distant seed brokers. In Vietnam, the Research Institute for Aquaculture, which is the leading government institute in fish breeding research and broodstock quality maintenance, has maintained public-private sector collaboration through national broodstock centers, which distribute tilapia broodstock to provincial centers, which multiply broodstock for further distribution to private hatcheries and fish farmers. There have been several tilapia seed production and distribution agreements in Bangladesh between the government-run Bangladesh Fisheries Research Institute and various private hatcheries, as well as among private enterprises. However, such collaborations are still in their infancy, with minimal technical and commercial interaction. The recent introduction of GIFT into Bangladesh from Thailand has piqued the interest of private tilapia hatcheries. In the

Asia-Pacific area, the development and distribution of GIFT and GIFT-related approaches have significantly increased public investment in the application of genetics to aquaculture in general, and tilapia farming in particular. The creation of national tilapia breeding programs, as well as staff and facility expansion, provides ample evidence for this. In Bangladesh, the Philippines, Thailand, and Vietnam, expanding freshwater aquaculture in general and tilapia farming, in particular, has become one of the key pillars for improving national fish supplies.

In the Philippines and Vietnam, the feasibility of tilapia exports has not been properly investigated. For tilapia producers to enter export markets, increased productivity and lower production costs are insufficient. When it comes to gaining access to tilapia export markets, the types and variety of products, packaging, promotional efforts, pricing, and distribution methods are all important things to consider. Tilapia is largely imported into the United States as frozen fish, frozen fillets, and fresh fillets, with fresh and frozen fillets primarily sold to retail grocers and restaurants through wholesalers and distributors. Live tilapia is extremely impossible to sell as food in export markets. To meet export demands, tilapia food products require post-harvest processing and value addition. This is much beyond the capabilities of any single farmer. As a result, entrepreneurship growth outside of the farm is critical. The importance of fish processing businesses and market intermediaries in exports cannot be overstated.

The vast and growing prevalence of GIFT and GIFT-derived strains in tilapia seed supply demonstrates their significant impact on farmed tilapia output. The usage of GIFT and GIFT-derived strains in the Philippines and Thailand is likely to continue, as these strains have emerged as the most preferred options among hatchery respondents in these countries for seed production in the next five years. Sex-reversed GIFT seed contributed an estimated 17% of national farmed tilapia production in Vietnam, and the overall contribution of GIFT and GIFT-derived strains to the national supply of Nile tilapia seed in Vietnam is expected to increase significantly as a result of the GIFT-based national tilapia breeding program(Kamaruzzaman, Nguyen, Hamzah, & Ponzoni, 2009). GIFT has yet to make a significant contribution to Bangladesh's national freshwater aquaculture production, but this is likely to change as farmed tilapia become more widely available and popular. Hatchery owners and grow-out farmers find tilapia farming to be a lucrative business. Hatchery surveys in the Philippines and Thailand revealed that hatchery operations were profitable. GIFT and GIFT-derived hatchery operations in the Philippines made great returns between 2001 and

2003, but faced tough competition and growing seed production expenses. The size of the hatchery was proven to improve respondents' optimism about their future activities. This emphasizes the need for economies of scale in tilapia seed manufacturing. GIFT hatchery respondents in Thailand enjoyed solid profitability from 2001 to 2003, owing to continued demand for GIFT and consistent marketing efforts. Due to a growing preference for high-quality seed with known provenance, they were confident about increased profitability over the next five years. According to available data in Vietnam, tilapia was more economical than other frequently accessible freshwater fish. Tilapia are currently in short supply, and many people are unfamiliar with the fish. The domestic market will be expanded by the expanding popularity and expansion of tilapia farming, increased awareness, and growing acceptability of tilapia among consumers, all of which will be aided by the wider availability of GIFT. In Bangladesh, the average annual tilapia consumption was around 2.5 kg/person in selected inland areas, but there was no data on average per capita consumption across the country(N. Ahmed, Young, Dey, & Muir, 2012). Many small-scale farmers have been discovered stocking tilapia in polyculture with carps in small ditches and borrow pits close to their homes. Tilapia is the most common fish consumed by the poorest households.

The production and distribution of GIFT naturally sparked concerns about whether they, and any changes to farming methods that they may bring about, would have negative consequences for the natural environment and biodiversity. Prior to and during the TA, great efforts were taken to identify and mitigate any potential negative consequences. Fish introductions and transfers, as well as success in fish genetic improvement, require an assessment of environmental repercussions, according to the researchers and creators of GIFT. The anticipated environmental implications of GIFT distribution were assessed further, and guidelines for the responsible introduction and use of GIFT were developed. The result was that the responsible development and diffusion of GIFT would not have a significant negative impact on the environment. Despite the developers of GIFT genetic resources' continued efforts to discourage the introduction of GIFT to African waters, where they would likely interbreed with and compromise the genetic integrity of important wild tilapia genetic resources, there are indications that the Asian private sector has already begun to sell GIFT to some African countries, and indeed to any country interested in buying GIFT. There are valid concerns about benefit sharing here. Africa provided Nile tilapia genes for GIFT creation and now wants to share the benefits of GIFT research and development. However, rather than exporting GIFT straight from Asia to Africa, such benefit sharing can be addressed by spreading and supporting the application of GIFT methodologies to new tilapia breeding initiatives in Africa that are based on African strains. The World Fish Center is working on transferring GIFT-related technology from Asia to Africa for use with native African tilapia, with UNDP funding. Given the continued attitudes toward controls and safeguards for much of the world's international trade in live fish, both farmed and aquarium species, GIFT and GIFT-derived strains are likely to be sought out and tested in African and other tilapia farming countries without adequate prior assessment of environmental impacts. Tilapia farming does not have a negative influence on the environment. Poor fish husbandry practices, as with any type of fish farming, can, however, lead to water pollution.

4.8 GENETIC IMPROVEMENT

Rapid genetic improvements can be accomplished in tilapia, which attain maturity at around 6 months of age. However, it took at least ten years for GIFT to be developed and implemented in national tilapia breeding programs. The quick proliferation of improved tilapia breeds had a significant impact in the countries that participated in this project. Improved tilapia breeds and long-term national tilapia breeding initiatives can dramatically and quickly increase tilapia farm yields and productivity. To cover the broad spectrum of germplasm collection, application of breeding methods and other protocols, on-station and on-farm trials, consolidation of research results, establishment of networks and national fish breeding programs, and distribution of improved breeds to hatcheries and fish farmers, the GIFT experience shows that substantial external TA and sustained funding were required for at least 10 years in the Asia and Pacific region. Furthermore, encouraging the use of enhanced tilapia breeds for farm productivity necessitated additional access to and investments in farm financing, support services, infrastructure, and marketing networks. The ADB's TA has had a high level of leverage, resulting in significant long-term positive outcomes and impacts on capacity development, knowledge and technology transfer, as well as policies and socioeconomic conditions in terms of tilapia production, farmer incomes, job creation, and human nutrition. This development experience, as well as the context in which the ADB pioneered, designed, and delivered the TA in collaboration with ICLARM and with widespread international backing, should be held up as a model for others to follow.

In many developing countries, biosafety, quarantine, and other environmental precautions are still being neglected when it comes to aquaculture development and research. The transmission of diseases and parasites, international transfers and domestic distribution of alien species and genetically modified farmed fish among and within such countries can put entire aquaculture industries in danger. Natural aquatic environments and biodiversity are also under risk, due to diseases and parasites, as well as predation, competition for food and spawning places, hybridization, and habitat change, among other things. To present, the production and spread of GIFT, GIFT-derived, and other Nile tilapia appear to have had no substantial negative effects on existing aquaculture or the natural environment and biodiversity in Asia and the Pacific. The region, however, is rich in freshwater biodiversity and habitats, and adequate portions containing this natural heritage should be preserved off limits to aquaculture, whether of local or foreign species, including tilapia, whenever practicable. Such places would function as in situ gene banks for future breeding programs of Asian farmed fish, containing wild genetic resources for future breeding projects(Nguyen, 2016). This could be accomplished by combining proper aquaculture development policies, rules, and regulations, as well as implementation, management, and administration techniques, with long-term aquatic biodiversity and habitat conservation.

Improved tilapia breeds and their transmission to farmers can be hampered by institutional, technological, and socioeconomic restrictions. It is insufficient to have strong research institutes for tilapia genetics and breeding when there are insufficient intermediary channels for supplying products and services to farmers as end-users. Strengthening ties between research and dissemination, as well as providing an enabling environment for tilapia farming to thrive, are both critical to the increase of farmed tilapia output. Future efforts should focus on public-private partnerships in tilapia research and information exchange, commercial alliances and partnerships in seed production and distribution to meet rising domestic demand for tilapia seed, and tilapia marketing to meet rising domestic demand for fish and export potential. To avoid potential conflicts of interest between the public and private sectors, concerned parties must clearly define and adhere to their respective roles, ensuring that the public sector does not suffocate the private sector's interests and that fish farmers, including the poor, have access to high-quality, low-cost tilapia seed. Interregional collaboration between African countries with the world's most important wild tilapia genetic resources but a scarcity of trained fish geneticists and breeding facilities, and Asian and other countries that are now leaders in tilapia farming and breeding. The availability of wild tilapia genetic resources from Africa and farmed tilapia strains from Asia was critical in the production of GIFT. This recommends that tilapia genetic improvement and breeding for tilapia farming should be pursued in tandem with tilapia genetic resource conservation in situ in African waters and ex situ on farms, in broodstocks, and in gene banks in all locations.

CHAPTER 5

ECONOMICS OF TILAPIA FARMING IN BANGLADESH

CONTENTS

5.1 INTRODUCTION

Bangladesh is considered one of the most appropriate countries in the world for freshwater aquaculture. In Bangladesh, aquaculture is growing faster than any other agricultural sector. Bangladesh is rated fifth in the world for aquaculture production, according to FAO statistics from 2016 (Hazrat Ali, Rico, Murshed-e-Jahan, & Belton, 2016)including homestead ponds, carps, tilapias, koi fish, shrimps, shrimps and prawns, prawns, rice and fish, and pangas. Farm type and farm owner characteristics were used as independent variables to explain observed chemical use. Forty-six chemical and biological products (7 water and sediment treatment compounds, 13 disinfectants, 7 antibiotics, 7 pesticides, 8 fertilizers and 4 feed additives and probiotics. In 2016-17, Bangladesh's total annual fish production was predicted to be 41.34 lakh MT, with 23.33 lakh MT coming from inland aquaculture, 11.63 lakh MT coming from inland capture fisheries, and 6.37 lakh MT coming from marine fisheries (Shamsuzzaman, Hoque Mozumder, Mitu, Ahamad, & Bhyuian, 2020). In 2016-17, the target for fish production was 40.50 lakh MT, but Bangladesh exceeded it by producing 41.34 lakh MT (Bogard et al., 2017). With this astounding achievement, Bangladesh became the first country in history to be self-sufficient in fish production, (Figure 5.1) supplying 62.8 g of fish per person per day in dietary consumption (B. Belton et al., 2011).

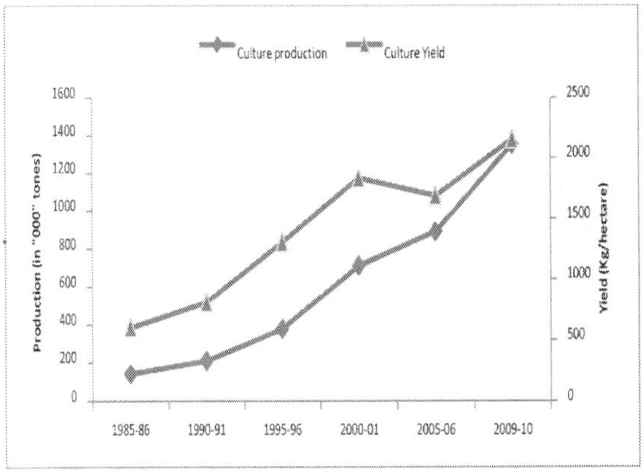

Figure 5.1: Bangladesh aquaculture.

Source: https://www.researchgate.net/profile/Md-Khan-65/publication/306285448/figure/fig6/AS:884773329326085@1587957808639/Production-trend-and-yield-of-aquaculture-in-Bangladesh-Source-DoF.png

Bangladesh is now one of the world's major fish producers, with the fishing industry accounting for 3.61 percent of GDP and 24.41 percent of agricultural GDP (Karim, Ahmed, Talukder, Taslim, & Rahman, 2006). Bangladesh has hundreds of seasonal water bodies in the form of ponds, ditches, small roadside canals, and barrow pits, all of which have enormous aquaculture potential. These are particularly well suited to the production of fish species with a short life cycle, rapid growth rate, and low input requirements. In such situations, tilapia can be a potential aquaculture fish in appropriate seasonal water bodies. In recent years, tilapia culture has spread throughout Bangladesh, with the country now ranking second in terms of annual pond fish production. Only tilapia accounts for 16.72 percent of pond fish output, whilst main carps like Rui, Catla, and Mrigal account for 29.56 percent (Md. A. Alam, Guttormsen, & Roll, 2019). The country's unusual riverine catfish farming has recently been badly damaged by low market prices. As a result, a huge number of commercial catfish growers have turned to tilapia as an alternative species to cultivate in their farms in order to increase profits.

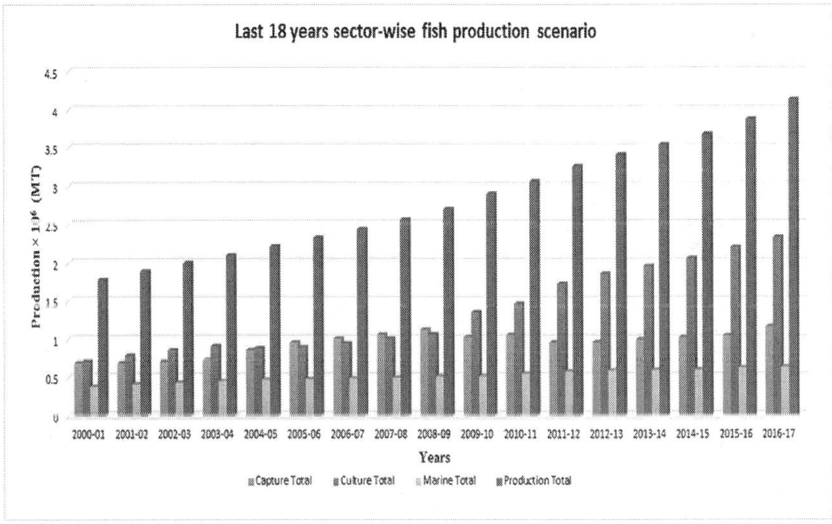

Figure 5.2: Fish export.

Source: https://ars.els-cdn.com/content/image/1-s2.0-S2468550X20300010-gr1.jpg

5.2 MARKET

Tilapia and Pangasius have a lot of potential in both the export and domestic markets (Figure 5.2). Many economic studies have been conducted on fish farming, but only a few have been conducted on tilapia farming in Bangladesh. Although findings from experimental stations are available, there are few studies on the economics of tilapia production at the farm level in Bangladesh. However, because these estimations were derived under controlled experimental conditions, they are not applicable to farm-level conditions. Most critically, the nature of tilapia farmers' reactivity to changes in input and product pricing is unknown. This information is critical because Bangladeshi farmers must not only be more efficient in their production activities, but also be more responsive to market indicators in order to maximize productivity and profitability in order to ensure supply to the urban market and improve farmer welfare. The impact of prices and fixed factors on farmers' resource allocation decisions was investigated using a profit function approach. This is due to a duality relationship between the profit function and the underlying production function. A profit function model has the advantage of being stated as a function of prices and fixed components, which are exogenous in origin and thus do not have the endogeneity problem that a production function model does. The fundamental assumption is that farm management decisions are a static profit maximization problem. The farm household is supposed to maximize restricted profits from growing specific fisheries, defined as the gross value of output variable expenses, under the constraints of a given technology and fixed factor endowments. The trans-log function, a flexible functional form that approximates most of the underlying real technology, was utilized (Madan M. Dey, Alam, & Bose, 2010).

The fish farmers interviewed faced a number of multifaceted obstacles and constraints like economic, social, and technological that had an impact on their fish farming activities and livelihood. Farmers were asked for their thoughts on issues that were found during data gathering. Polyculture of fish is a typical pond-based aquaculture technology used by small-scale producers in developing nations to maximize productivity and revenues by improving input use efficiency. Polyculture is the most common system in these ponds, with 66 percent of them using it. These households' questionnaires were performed in 2019 (Kabir & Eva, 2014). Farmers using tilapia polyculture systems, on the other hand, conserve 72 kg each cycle more farmed fish for household use than farmers of other systems (Debnath et al., 2022). After correcting for country-level characteristics, the greater profits for general polyculture, tilapia polyculture, and rohu polyculture are statistically significant, with a high critical value of gamma in the Rosenbaum sensitivity analysis, showing that these results are resilient. This examination of fish farming households adds to the findings of pond experiments and can aid in aquaculture policy and training decision-making.

5.3 AQUACULTURE

Aquaculture contributes greatly to household income and nutrition, as well as local and nationa l economies. At a worldwide level, a critical transition is occurring as capture fisheries give way to aquaculture. While capture fisheries will remain significant, aquaculture has demonstrated its importance in maintaining global food and nutrition security. Inland aquaculture with freshwater fish contributes the most to total aquaculture volumes, rural livelihoods, and food security. Aquaculture contributes to a 2.1 percent rise in income in Bangladesh and lifts 18 million people out of poverty (Toufique & Belton, 2014). By increasing fish consumption, aquaculture revenue and production can help fish farming households improve their food and nutrition security. Tilapia and rohu carp polyculture were studied because they are the two most commonly farmed fish among the farmers. Fish farming tends to concentrate on high-value species at this stage, such as Indian big carps, of which rohu is the most popular. Rohu carp cultivation has a high consumer demand and expansion potential, making it one of the most significant farmed fish species in India and neighboring countries including Bangladesh, Pakistan, and Myanmar. The low-value species are rising in popularity in Bangladesh, with tilapia being devoured more than rohu carp. Tilapia was first brought to poor countries, where it was farmed on a subsistence level to supply protein requirements. After carps, tilapia has risen to become the second most significant farmed fish, with Indonesia, China, and Egypt as the leading producers. WorldFish conducted the surveys in 2019, and the dataset includes 1,307 mostly small-scale fish farming families with 1,651 ponds, 73 percent of which produce tilapia and 53 percent of which produce rohu carp (B. N. Ahmed & Waibel, 2019). Producers in the three countries have distinct tastes for farmed fish. The comparison of fish farming systems tends to focus solely on on-farm performance measures, with less attention paid to other factors such as fish farming households' food and nutrition security. We can incorporate the outcome variable of harvest saved for household use and find that polyculture farmers, particularly tilapia polyculture farmers, keep more of their farmed fish for their personal consumption. This is critical for food security because fish is a good source of macro- and micronutrients. Consumption of fish may not give a complete picture because these households may supplement their catch with fish purchased from the market. When more extensive food consumption data for fish farming households becomes available, it will be possible to examine how different aquaculture methods affect food and nutrition security.

Despite the fact that Bangladesh is one of the best countries in the world for freshwater aquaculture, tilapia farming is not widely practiced. Despite the fact that tilapia has been introduced to Bangladesh since 1954, the culture of the fish has yet to take root due to social, technological, institutional, and marketing obstacles. While tilapia farming has enormous potential, its production, financial feasibility, and long-term viability are all dependent on market conditions. Market factors play a big role in determining the profitability of tilapia farming. Studies were carried out to better understand the current practices of tilapia marketing systems in Bangladesh in order to ensure the country's long-term food security. Primary data was obtained through a field survey that lasted nine months and included participatory rural appraisal, quick market appraisal, questionnaire interviews, and cross-check interviews with key informants. To better understand tilapia harvesting and selling systems, data was collected in the Trishal Upazila that is a sub-district of the Mymensingh district in north-central Bangladesh. Almost all tilapia produced in Bangladesh is sold within the country for local use. From farmers to consumers, the tilapia market chain consists mostly of primary, secondary, and retail markets, involving local agents, suppliers, wholesalers, and retailers. Despite the fact that the tilapia marketing system is conventional and less competitive, it is critical in connecting farmers and consumers, and thus contributes significantly to the value-added process. Market demand for tilapia is great, but supply is limited, resulting in the formation of a robust network of intermediaries and dealers who act as a link between farmers and consumers. Tilapia prices are determined by their quality, size, and weight, as well as supply and demand, the availability of other fish species in markets, market location, market infrastructure, and seasonality. Despite the potential of tilapia marketing, traders reported a number of challenges, including poor road and transportation infrastructure, higher transportation costs, insufficient ice supply, a lack of credit, poor market infrastructure and unsanitary conditions, poor institutional support, and inadequate training facilities.Despite the fact that Bangladesh is one of the best countries in the world for freshwater aquaculture, tilapia farming is not widely practiced. Despite the fact that tilapia has been introduced to Bangladesh since 1954, the culture of the fish has yet to take root due to social, technological, institutional, and marketing obstacles. While tilapia farming has enormous potential, its production, financial feasibility, and long-term viability are all dependent on market conditions. Market factors play a big role in determining the profitability of tilapia farming. Studies were carried out to better understand the current practices of tilapia marketing systems in Bangladesh in order to ensure the country's long-term food security. Primary data was obtained through a field survey that lasted nine months and included participatory rural appraisal, quick market appraisal, questionnaire interviews, and cross-check interviews with key informants. To better understand tilapia harvesting and selling systems, data was collected in the Trishal Upazila that is a sub-district of the Mymensingh district in north-central Bangladesh. Almost all tilapia produced in Bangladesh is sold within the country for local use. From farmers to consumers, the tilapia market chain consists mostly of primary, secondary, and retail markets, involving local agents, suppliers, wholesalers, and retailers. Despite the fact that the tilapia marketing system is conventional and less competitive, it is critical in connecting farmers and consumers, and thus contributes significantly to the value-added process. Market demand for tilapia is great, but supply is limited, resulting in the formation of a robust network of intermediaries and dealers who act as a link between farmers and consumers. Tilapia prices are determined by their quality, size, and weight, as well as supply and demand, the availability of other fish species in markets, market location, market infrastructure, and seasonality. Despite the potential of tilapia marketing, traders reported a number of challenges, including poor road and transportation infrastructure, higher transportation costs, insufficient ice supply, a lack of credit, poor market infrastructure and unsanitary conditions, poor institutional support, and inadequate training facilities.

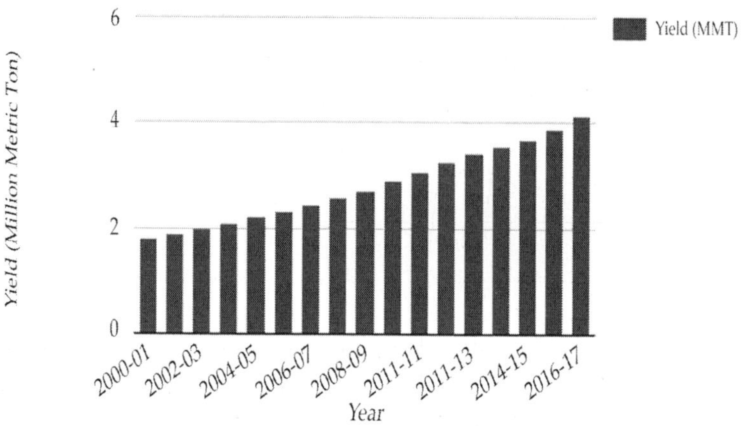

Figure 5.3: Fish demand.

Source: https://www.lightcastlebd.com/wp-content/uploads/2019/04/1-6.png

5.4 DEMAND

Bangladesh is one among poor and most densely inhabited countries on the planet (Figure 5.3). Rice and fish are the most significant food crops for Bangladesh's 140 million inhabitants (Saiful Islam, Barman, & Murshed-e-Jahan, 2015)different national and international organizations are actively promoting integrated rice–fish farming system (IRFFS. Fish are key sources of protein, critical vitamins, minerals, and fatty acids for the Bangladeshi population. Fish accounts for almost 70% of animal protein intake, with an average yearly fish consumption of 14 kg per person. Fish consumption per capita is lower than the global average of 16.1 kg per year (Dasgupta, Mustafa, Paul, & Wheeler, 2021). Nonetheless, Bangladesh has long recognized the value of aquaculture as a food supply. Because of its favorable agro-climatic conditions, Bangladesh is considered one of the best countries in the world for freshwater aquaculture. The best circumstances for fish production include a subtropical temperature and wide regions of shallow water. In 2006, total annual fish production was predicted to reach 2.32 million tonnes, with inland aquaculture accounting for 38 percent, inland capture fisheries accounting for 41 percent, and sea fisheries accounting for 479,810 tonnes (Zaman, Jewel, & Bhuiyan, 2006). In Bangladesh, the principal freshwater aquaculture production systems are vast and semi-intensive pond polyculture of Indian major carps and foreign carps, which account for 80% of total freshwater aquaculture production.

There is a significant potential for tilapia farming in Bangladesh to address the rising need for food. Due to its rapid development rate, high market demand, and increasing consumer acceptance, tilapia aquaculture is widely practiced in several Asian nations, including China, Indonesia, the Philippines, Thailand, and Vietnam. Tilapia have overtaken carps as the world's second most popular farmed fish due to their growing appeal among consumers. In Bangladesh, tilapia farming has a long history, and it was once thought that tilapia would be a miracle fish in aquaculture.

Figure 5.4: Oreochromis mossambicus.

Source: https://www.fishbase.se/images/species/Ormos_m0.jpg

In 1954, the Mozambique tilapia or *Oreochromis mossambicus* (Figure 5.4) was imported from Thailand to Bangladesh. However, because of its early maturation and prolific breeding, this species was not widely adopted for aquaculture due to pond overcrowding. Despite its long history of introduction to Bangladesh, tilapia culture has been unable to establish itself in aquaculture due to socioeconomic, technological, institutional, and marketing obstacles. Nonetheless, due to its profitability, a large number of farmers in rural Bangladesh have become involved in tilapia production in recent years. GIFT has become a popular fish among farmers among various tilapia kinds. Under subsistence fish farming methods that allow for a minimum of two crops per year, this fish reaches marketable size in four months. Even though tilapia farming has yet to make a considerable contribution to national freshwater aquaculture production, this is anticipated to change as farmed tilapia become more widely available and popular.

While tilapia farming in Bangladesh has a lot of potential, its production, commercial feasibility, and long-term viability are all dependent on markets. Market factors play a big role in determining the profitability of tilapia farming. The tilapia market is characterized by high demand, which is fueled by population growth. There is a widening disparity between supply and demand for tilapia in markets, attributable primarily to population expansion. To close the gap, not only must tilapia production be increased, but all aspects of marketing and distribution networks must be improved. It is thus worthwhile to establish tilapia markets in order to ensure its long-term viability. Apart from gaining a better understanding of tilapia marketing systems, it appears that identifying marketing inefficiencies that negatively affect poor farmers, merchants, and related groups, as well as tilapia output, is critical. Since 1995, along the Dhaka-Mymensingh corridor, fish aquaculture has exploded in the Mymensingh area. Due to the decreasing market price of pangas, several farmers have recently converted to tilapia farming. Tilapia is grown in a variety of settings, including small-scale, low-input rural ponds, semi-intensive, intensive, and commercial enterprises. In the Mymensingh area, the average size of a fish farm was observed to be 0.23 hectares (H. Ali & Haque, 2011). The majority of farmers practice small-scale tilapia farming with carps and catfish, while only a few farmers and business people cultivate tilapia as a monoculture in semi-intensive or intensive systems. The peak season for tilapia farming is from April to December, which is a nine-month culture period. Farmers start stocking their ponds in April or May and harvest tilapia four months later, usually twice a year. Tilapia farming is entirely reliant on hatchery-produced fry.

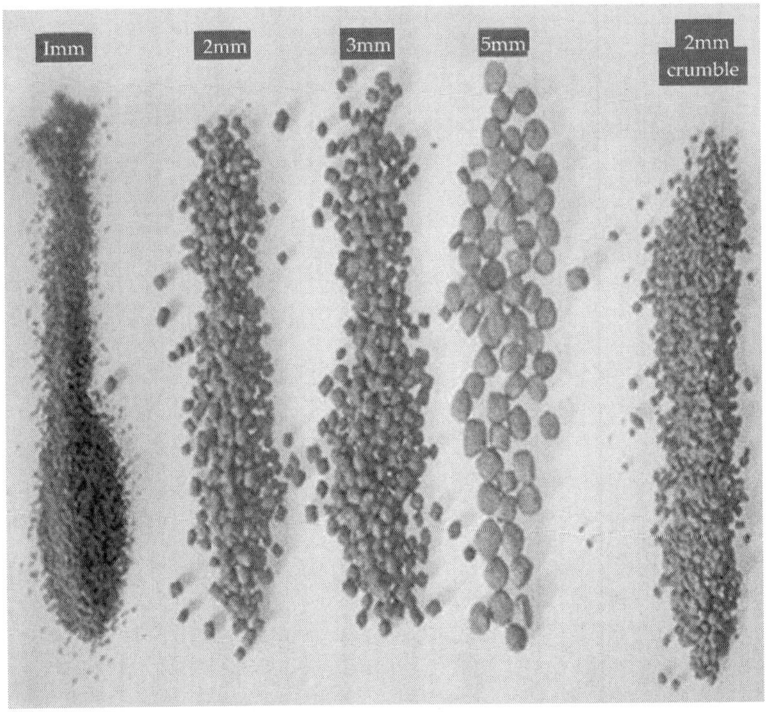

Figure 5.4: Tilapia feed.

Source: https://www.fish-feed-extruder.com/wp-content/uploads/2017/09/feed-size.jpg

5.5 FEED

A blend of rice bran, wheat bran, and mustard oil cake is the most frequent supplementary feed for small-scale tilapia farming, and it is widely available on-farm or in local markets (Figure 5.4). Nonetheless, industrially made pelleted feed remains the mainstay of intense or semi-intensive tilapia aquaculture. Feed is usually provided twice a day, in the morning and in the evening. For tilapia grow-out, farmers employ a variety of fertilizers, including cow dung, urea, and triple super phosphate. Farmers' households consume a lesser percentage of their harvested tilapia, and they prefer to eat smaller fish rather than sell them. Farmers instead sell larger tilapia to fund the acquisition of low-value fish like pangas and silver carp. In comparison to carp polyculture, tilapia farming faces a number of challenges, including a difficult production system, a lack of technical support, limited availability and poor-quality seed, high seed and feed prices, low tilapia market prices, poor marketing facilities, and lower economic returns. Notwithstanding the fact that a large number of poor farmers engage in tilapia farming, they are unable to participate in semi-intensive or intensive culture systems due to a lack of financial support. Uncontrolled tilapia breeding in ponds is also an issue, however, some farmers perceive this breeding as a benefit because it provides seed.

5.6 APPRAISAL

Rapid Market Appraisal (RMA) is a quick and easy way to get policy-relevant and intervention-oriented data on any commodity sub-sector. RMA can be used to identify the tilapia marketing channel, visit physical facilities such as tilapia landing sites and markets, and observe tilapia trade operations directly. Several visits to tilapia production areas and different markets were performed for research in order to observe tilapia distribution and marketing systems. Interviews were directed using a short questionnaire that covered tilapia marketing systems and their limitations. This includes preparation and testing of the questionnaire, application of statistical processes to identify the optimum sample size and sampling method, and training of enumerators to fill out surveys all preceded by questionnaire interviews with tilapia traders. Reconnaissance for the pilot survey and survey method improvement were among the pre-survey activities. Primary interviews with vendors were done in order to prepare the questionnaire. Any new information that had not been planned to be asked was incorporated. Based on the results of the pilot survey, the questionnaire can be tweaked and enhanced. Tilapia traders were chosen for questionnaire interviews using simple random sampling. Because most tilapia farmers and traders were unaware of the research, it took time for the researchers to persuade them. A substantial volume of time and effort had to be spent persuading respondents of the survey's credibility.

In the beginning, some of the fish traders were uninterested in participating in the interview. Despite being presented with an identity card prior to the interview, they suspected the researchers of being employees or agents of the tax office, police department, or other official agency. Respondents' reactions to some of the questions related to costs, returns, and profits indicated that they would not be willing to speak freely unless they were entirely confident in the interviewers. During this time, more effort was put into persuading the respondents of the researchers' and the study's credibility. Some respondents were dubious of the researchers' identities and refused to speak with them until government fisheries inspectors gave them permission. This primarily affected persons without a formal education. When possible, the Department of Fisheries (DOF) was enlisted to assist in the interviews of these individuals. Although the tilapia farmers' and merchants' sentiments and comments were generally favorable, this was not always the case when they were queried about costs, returns, or incomes. Financial data that could be trusted was extremely difficult. The majority of farmers, traders, and related groups lacked basic accounting notes. Even if some farmers and traders gave financial information, it was based on guesswork that may or may not be true.

Figure 5.5: Tilapia harvesting.

Source: https://www.fao.org/typo3temp/pics/a1076dc1d0.jpg

5.7 HARVESTING

Tilapia harvesting (Figure 5.5) begins as soon as the fish reach marketable size. Although a few major farmers rely on commercial harvesters, most farmers gather tilapia themselves. Tilapias are typically collected early in the morning. The majority of tilapia farmers use partial harvesting, which permits smaller fish to develop, but a few farmers use total harvesting. Farmers catch tilapia with cast nets and seine nets, netting numerous times over the course of a few weeks. Tilapias are harvested, cleansed with pond water, and stored in aluminum canisters or bamboo baskets until sold. Farmers sell their produced tilapia to local agents or suppliers, depending on transaction volumes. Local agents purchase tilapia from farms near the pond and transport them to vendors. Farmers, on the other hand, transport tilapia to suppliers from far away areas. Despite significant advances in road infrastructure, isolated locations still confront an accessibility issue, which has an impact on tilapia marketing. Heavy rains frequently ruin village roads, rendering rickshaws, trucks, and motorized vehicles nearly immobile on the muddy roads of the villages. As a result, transportation expenses are high, and profit margins are low. In addition to these issues, farmers are in a particularly poor position i.e., no price negotiation power in relation to tilapia selling middlemen. According to the report, high transportation costs are the single most important limitation for tilapia marketing for respondents. Poor road and transportation infrastructure was identified by the respondents. Only some farmers cited low market prices as the most significant limitation. Because production decisions such as investments and profits are based on current market prices, any market decline will have an impact on the profitability and feasibility of tilapia farming.

Almost all tilapia produced in Bangladesh is sold within the country for local use. The tilapia market is limited in terms of volume, value, and employment. The tilapia marketing chain employs a large number of people, many of whom live in poverty, as farmers, suppliers, traders, transporters, and day laborers, including women and children. The tilapia marketing system is conventional and less competitive, but it is crucial in connecting farmers and consumers, and hence contributes significantly to the value-added process. In tilapia marketing systems, farmers are the primary producers. Farmers, with a few exceptions, never contact consumers directly with consumers. From farmers to consumers, the tilapia market chain consists mostly of primary, secondary, and retail markets, involving local agents, suppliers, wholesalers, and retailers. Market demand for tilapia is great, but supply is limited, resulting in the formation of a robust network of intermediaries and dealers who act as a link between farmers and consumers. A handful of strong intermediaries manages, finances, and controls practically all aspects of tilapia marketing. Wholesalers are the most influential middleman. Supplier-to-wholesaler communication is often good and takes place using mobile phones. Suppliers are a type of middleman who transport tilapia from primary markets to wholesale markets. Suppliers are often related to a small number of distributors. A provider transported an average of 18 kg of tilapia per day, ranging from 10 to 39 kg per day. Tilapia is widely transported to wholesale markets in Mymensingh and Dhaka by trucks, buses, pickup trucks, and taxis, which takes 1 to 4 hours depending on distance and mode of conveyance (N. Ahmed, Ahammed, & Lecouffe, 2007)where the livelihoods of a considerable number of people are associated with prawn distribution and marketing systems. The market chain from producers to consumers passes through a number of intermediaries. The prawn is a highly valued product in international markets; most of the prawns (60%.

The wholesalers handle the landing, handling, and auctioning of tilapia by size groups as soon as the suppliers land tilapia in the wholesale market. The wholesalers employ a number of day laborers. Cleaning, sorting, grading, and frosting fish are among the jobs they conduct. Normally, piles are used to sell items at an auction. Wholesalers adhere to the incremental pricing method. It is the most competitive type of auction, resulting in higher prices. The majority of sales agreements are informal, which means that the price of tilapia, quantity, and other trading terms are decided orally. Their business and product transactions are mostly based on mutual trust. A small number of wholesalers are also related to the retailers. Retailers and wholesalers have a usually positive connection. The selling of tilapia necessitates a significant sum of money. Wholesalers in Dhaka marketplaces have greater capital than wholesalers in Mymensingh markets. The majority of funding for tilapia marketing comes from a variety of personal and informal sources, as well as loans. Tilapia marketing is mostly funded by wholesalers selling household assets. Some people have their own money, either from savings or from the sale of personal assets such as gold jewelry, livestock, timber, and land. Despite the fact that the real sale of land to finance fish marketing appears to be extremely rare these days, land leasing is quite widespread.

Many wholesalers are unable to fund tilapia marketing and must rely on a variety of financial sources. Several organizations, including banks and local moneylenders, have emerged in recent years to provide loans for fish marketing. One of the most critical aspects driving tilapia marketing is financial availability. Moneylenders provided 78 percent of wholesalers with credit, while banks provided 22 percent (Paprocki & Cons, 2014)it explores the class-differentiated implications of food sovereignty in a zone of ecological crisis – Bangladesh's coastal Khulna district. Much land in this deltaic zone that had previously been employed for various forms of peasant production has been transformed by the introduction of brackish-water shrimp aquaculture. This has, in turn, caused massive depeasantization and ecological crisis throughout the region. Through an examination of two markedly different polders (embanked islands. In Mymensingh marketplaces, more wholesalers were given credit than in Dhaka markets. Moneylenders are the most popular source of credit for tilapia marketing. Local businessmen, on the whole, act as moneylenders. Moneylenders and wholesalers often have a strong working relationship. A moneylender's typical monthly interest rate is ten percent. Wholesalers can also get loans from local branches of major banks, with land as security and at an annual interest rate. Wholesalers who received bank loans, on the other hand, reported that they would prefer not to go to banks because of the excessive paperwork and limited loan amounts. Tilapia trade has become a viable industry for its participants over time, as well as a source of new jobs. The potential for big earnings has lured the majority of traders to tilapia trading. However, because tilapia availability is irregular, traders also engage in other types of fish trading. The peak season for tilapia marketing is from September through December, despite the fact that tilapia trading is a year-round activity. Markets are open every day, and because of the increased supply of tilapia.

Rental of the market area, energy, labor, ice, transportation, and miscellaneous are all expenses associated with tilapia marketing. The volume of tilapia, the number of laborers necessary, market distance, mode of transportation, and market infrastructure all influence marketing expenses. The typical marketing expenditures in Kawran Bazar were higher than in other retail markets, owing to increased prices for energy, ice, transportation, and labor. On the other hand, Machua Bazaar had lower average marketing expenditures due to reduced costs for all items except miscellaneous. These apparent variations in marketing expenditures, however, were not significant. The multiple regression of tilapia price versus supply was negative but significant at the 1% level, indicating that an increase in tilapia supply would lower market price and a reduction in tilapia supply would raise the market price. The price of tilapia is influenced by a number of factors, including supply and demand. The biological environment, the technology utilized, the policy and institutional environment, and the producer's profile all influence tilapia supply. Similarly, consumer demand is determined by the policy and customer profile. Changes in any of these variables affect tilapia production, demand, trade, and prices. On the supply side, the price of tilapia is influenced by the seasonality of production, which affects the amount of goods accessible on the market. Tilapia prices rise when supplies are short. Inter-seasonal pricing changes are also influenced by demand behavior. Tilapia prices in the research area are normally lower between September and December, then rise over the next four to five months.

The price of tilapia and the weight of the fish had a favorable association. The apparent price differences between the various weights of tilapia were found to be statistically significant. Variables such as marketing costs and margins, the number of intermediaries in the marketing channel, the distance between primary and retail markets, and consumer price behavior are all significant considerations to consider when analyzing the value chain of tilapia marketing. In tilapia marketing systems, a large number of intermediaries are engaged. As the number of middlemen in the marketing route grows, farmers' profit margins shrink. The presence of intermediaries at various levels of the marketing channel is mostly owing to a lack of government regulation of the industry. The intermediaries take advantage of the situation and exploit both farmers at the farm gate and consumers at the retail level. In the absence of government interference in the market, the middlemen set the price for tilapia and decided on a considerable margin. Farmers are susceptible to negotiations with market entities with greater financial endowments, such as local agents, suppliers, wholesalers, and retailers. This situation is worsened by a dearth of transparency in the price formation process and asymmetric information flow, a lack of capital for investment, and insufficient postharvest infrastructure, particularly poor road and transportation facilities, as well as insufficient market services, all of which are essential for the marketing of high-quality products.

Despite the potential of tilapia marketing, traders reported a number of challenges, including poor road and transportation infrastructure, higher transportation costs, insufficient ice supply, unsanitary conditions, a lack of credit facilities, and poor market infrastructure i.e., inadequate drainage systems, poor water supply, limited ceiling and flooring space. Strikes and roadblocks, for example, have an impact on tilapia marketing. As a result, merchants sell tilapia at low prices, often losing money entirely. Frequent power outages can have an impact on ice manufacturing, which has an impact on the marketing of high-quality tilapia. There are currently no industry standards for tilapia handling, washing, sorting, grading, cleaning, or icing. Fish markets include rudimentary facilities, as well as low hygiene and sanitation. Most fish landing stations have inadequate support infrastructure and are thought to be poorly managed and unsanitary by some. In some secondary market centers and auctioning places, shelters are not available.

The majority of fish damage happens during transit from primary markets to retail outlets. In general, retail market conditions in terms of stalls, spacing, sanitation, drainage, and management are far from ideal. Quality control is rarely carried out at the landing, handling, distribution, and marketing locations. This is owing to a scarcity of quality control inspectors and a lack of attention to quality control for domestic markets.

5.8 SOCIOECONOMICS

The socioeconomic circumstances of traders' houses are crucial in the design of development operations. When family members have appropriate access to cash and resources to meet basic needs through fish selling operations, a household's social conditions can improve. Food, shelter, health facilities, drinking water, and sanitary facilities are examples of basic necessities. Most traders, however, live in substandard housing, which has an impact on tilapia marketing since traders are more inclined to spend available monetary resources on house maintenance rather than fish trading. Poor health facilities also have an impact on tilapia marketing because of labor shortages. Furthermore, rather than investing in tilapia trading, traders are more likely to spend money on household members' medication. Due to inadequate sanitary conditions, most traders indicated that members of their homes frequently suffered from diarrhea and cholera. As a result, the lack of hygienic facilities has an impact on not just the function of traders in tilapia selling, but also their wider income-generating potential. Tilapia traders from various markets have varying levels of education, age, family size, and income (Rand & Tarp, 2009).

Despite the evidence on tilapia demand, tilapia has significant food supply potential in Bangladesh. If tilapia can produce in large quantities, the marketing of tilapia in Bangladesh could help meet the rising demand for fish among consumers, including the poor. The growth of a sustainable tilapia market, as well as the widespread adoption of tilapia farming, would have a considerable impact on family food security. Tilapia consumption has been increasing at the household level. In Bangladesh, tilapia appears to be invited by all religious and social groups. Consumers like fresh tilapia that haven't been iced, according to the traders. In general, high-income groups such as government officials and wealthy businesses purchase huge tilapia, while the middle class can only afford medium and tiny fish. Due to equality concerns, a major part of the poor population buys little tilapia i.e., one tilapia per household member. Despite this, most customers favored larger tilapia owing to taste, although smaller fish were typically purchased due to cheaper market prices. Household consumers, on the whole, flock to retail marketplaces to buy fish. Due to its increasing popularity, restaurants, highway motels, guest houses, resorts, department stores and institutional consumers purchase tilapia. Residents of university halls, patients of medical hospitals, and participants of conferences, seminars, and training centers have all become significant outlets for aquaculture products, including tilapia. Tilapia is also served at parties and other events where guests are entertained. Most dealers indicated that tilapia is growing increasingly popular among all demographics, including women, children, the young, and the elderly, due to its delicious taste and lack of bones, making it easy to eat. Women frequently buy tilapia in Dhaka marketplaces as they tend to the kitchen and maintain household nutrition. Lunch, rather than breakfast or dinner, is the most common meal for moderate income groups and the poor when eating tilapia with rice. Nonetheless, high-income people prefer to eat tilapia at home or in restaurants for dinner.

While there is a high demand for tilapia, there are a number of obstacles to its marketing, including poor road and transportation facilities, higher transportation costs, insufficient ice supply, unsanitary conditions, a lack of credit facilities and poor market infrastructure, inadequate drainage systems, poor water supply, and limited ceiling and flooring space. Despite these drawbacks, tilapia marketing has a number of positive effects on the food supply. With Bangladesh's growing population and demand, expanding fish supplies to preserve food security has become a top issue. Because of several beneficial cultural qualities, there is a significant potential for tilapia farming in Bangladesh to meet the increasing demand for food. Small-scale farmers, on the other hand, have the potential to participate in aquaculture and gain from it when there is a market for fish. Market factors have a big impact on the profitability of tilapia farming. The findings reveal that tilapia farmers cannot make a profit because they have restricted access to the market and lack expertise about where to sell their products. Furthermore, due to exploitation by intermediaries, the farm-gate price of tilapia is low. The profit margins of framers in the tilapia marketing channel decrease as the number of middlemen increases. The presence of intermediaries in the marketing channel is owing primarily to a lack of government oversight. Farmers and customers are both exploited by intermediaries.

The Tilapia market is completely dominated by powerful intermediaries and benefits are only shared among a small number of people. Wholesalers are primarily responsible for facilitating the supply of tilapia from producers to customers via various intermediaries. Despite the fact that wholesalers play a major part in tilapia marketing, they have been designated as the most accountable factor for consumer price increases. As a result, wholesalers may be able to assist in balancing market pricing, which would benefit both producers and consumers. A sufficient supply of tilapia is required to keep markets afloat. With greater supply, the market for tilapia will grow. However, current tilapia production and availability are insufficient to meet market demand. Despite the fact that household tilapia output is still low due to lesser acceptance of tilapia farming, small pond sizes, and lower rates of production from low-input technology, household tilapia production is still low. More tilapia production and on-farm consumption can be directed with improved technology and management practices. The direct effects of tilapia eating on agricultural adoption can be enormous. Adoption of tilapia aquaculture may boost market supply, lowering prices and thereby increasing tilapia consumption.

Better marketing and transportation infrastructure for tilapia would assist to improve the issue. Improved fish marketing infrastructure, including market centers and facilities, is required to improve market access and sales mechanisms. To better serve consumers, efficient tilapia distribution channels should be established. To maintain a seamless supply of tilapia to customers, it is also worthwhile to create tight cooperation between farmers and market operators. Government institutions provide technical advice and support on marketing, market facility design, and transportation, incorporating NGOs where appropriate, and the implementation of a management plan to handle existing constraints, as well as capacity building for stakeholder organizations. Numerous strategies have been put in place to eliminate social and institutional barriers that prohibit more people from being involved in fish farming and selling. Overall, there aren't enough policies in place to ensure the long-term success of tilapia marketing. Integration of production systems, as well as the creation of backward or production inputs and forward or postharvest and value-added activities, linkages, can aid in tilapia marketing development. Through policy implementation, the public sector plays a vital role in developing tilapia marketing. Market planning, market infrastructure construction, and market maintenance are all handled by DOF. In order to promote sustainable tilapia marketing, a conceptual framework has been devised that comprises three main components: market awareness, market access, and marketing facilities, all of which are thought to be interconnected. Farmers and merchants of tilapia are more willing to participate in innovation as a result of this approach. Attempting an innovation, on the other hand, is highly dependent on the scope of market access. To have an innovation implemented, a fundamental paradigm shift would be required to determine market awareness, market access, and marketing capabilities of stakeholders. More research, proper extension efforts, support services, and public-private partnerships could have a big impact on tilapia marketing innovation.

Marketing tilapia in a sustainable manner can help to enhance food supply. Some studies found a number of barriers influencing tilapia marketing efficiency, necessitating the implementation of relevant solutions. Tilapia markets frequently lack essential infrastructure such as clean water, a sufficient drainage system, and ceiling and flooring. Poor road and transportation infrastructure is also a hindrance to tilapia marketing. It is consequently critical to upgrade fish landing, transportation, handling, and preservation facilities in order to ensure a steady supply of high-quality fish. Farmers, traders, and intermediaries appear to have a weak understanding of sanitary norms and fish quality. The cleanliness of the fish markets is also critical. It is necessary to maintain proper management in terms of day-to-day hygienic maintenance of the facilities. Improving the sanitary environment of fish, such as tilapia landing hubs and marketplaces, is critical for generating high-quality products. As a result, tilapia market operators should receive training in areas such as preservation, handling, icing, and curing. One of the most important challenges for fish preservation is a lack of ice in markets. Ice is essential for the storage and preservation of high-quality seafood. The availability of ice on the premises would aid in the improvement of proper tilapia handling. As a result, a sufficient number of ice factories must be established for the marketing of high-quality fish. Due to extensive government paperwork and collateral arrangements, farmers, traders, and allied groups do not have simple access to bank and NGO credits. As a result, farmers and merchants should be assisted in obtaining more affordable bank financing for tilapia production, distribution, and market operation costs. Given the scarcity of information services among producers, distributors, and market actors, as well as among development organizations, the creation of an information network is critical. Transporting, handling, icing, preservation, and curing training for tilapia market operators would increase product quality in marketplaces.

CHAPTER 6

TILAPIA CULTURE IN INDIA

CONTENTS

6.1 INTRODUCTION

Indonesia, Thailand, the Philippines, and Taiwan produce the most tilapia in the world. Tilapia is farmed in around 85 nations, with 98 percent of the tilapia produced in these countries being raised outside of their natural habitats (Behera et al., 2018). The Nile Tilapia or Nile Perch is the most widely farmed tilapia breed in the world, accounting for around 75% of all farmed tilapia (Raj et al., 2019). Tilapia can be cultivated in a range of aquaculture conditions, including brackish and saltwater, as well as pond and cage systems. Tilapia is only available during certain times of the year. Only warmer water allows it to survive and reproduce. A tilapia farm's optimal water temperature is between 82oF and 86oF (Pradhan et al., 2020). Fish will begin to die at temperatures below 55 degrees, and their growth rate will slow. As a result, India is an ideal location for tilapia aquaculture. Three months before they are ready to eat, tilapia begin reproducing. It has a very high rate of reproduction. A female tilapia fish can produce up to 100 fry every week as an adult (Macintosh, Varghese, & Rao Satyanarayana, 1985). To breed male tilapia, several breeders employ hormones to regulate the breeding process. In the grow-out stage, most farmers choose to keep only male tilapia. Male tilapia has proven to be more profitable as they grow larger and consume less time and energy. Due to breeding, female tilapia loses a lot of energy and time. Tilapia is the world's second most farmed fish, yet commercial tilapia farming is limited in India. Despite the fact that this fish was brought to India many years ago, and despite a restriction on tilapia farming by the Indian fisheries research committee in 1959, tilapia cultivation has lately been permitted in select states with certain criteria (Subrahmanyan, 1959). Farming of genetically enhanced tilapia (GIFT) was permitted with minor restrictions. In the late 1970s, the Nile Tilapia was introduced to India (Haridas et al., 2017). Nile Tilapia were found in small numbers in the river Yamuna in 2005, but their proportion climbed to roughly 3.5 percent of total fish species in the river in just two years (A. Alam et al., 2015).

The amount of tilapia in the Ganga River system is currently around 7% of the total fish species. The optimal temperature for tilapia growing in India is 15oC to 35oC. Tilapia, on the other hand, can survive in temperatures ranging from 10 to 40oC (Gupta, Dwivedi, & Tripathi, 2018). Because of their unconditioned multiplication, the main issue in tilapia farming is having varied ages and sizes of tilapia fish in the same pond. Because male tilapia grows quicker than female tilapia, growing male tilapia by isolating females is referred to as monosex tilapia fish farming. Commercial tilapia farming makes a lot of money because the male tilapia adapts well to additional feeding and grows quickly. In both domestic and foreign markets, this fish is in high demand. In commercial tilapia production, providing nutritious food is critical for the fish's rapid growth and increased body weight. Tilapia eats plants, especially protein-rich duckweed which has the same protein content as a commercial fish diet, and uses microscopic combs in its gills to filter algae from the water. Duckweed and commercial fish feed are good together, although tilapia can grow on duckweed alone. Diversification of species in aquaculture by introducing more species to increase output levels has become vital as the demand for fish grows. The addition of tilapia to our culture systems is helpful since it represents a lower level in the food chain, making its culture more cost-effective and environmentally friendly. The advantages of mono sex tilapia cultivation include faster growth and males that are larger and more uniform in size.

Figure 6.1: Mono sex tilapia.

Source:https://fishseedsuppliers.com/wp-content/uploads/2019/12/monosex-tilpia.jpg

6.2 MONO SEX TILAPIA

The mono sex tilapia fish (Figure 6.1) farming industry is growing at a rapid pace because of its high profitability. There are subsidies available for commercial tilapia fish farming if anyone wants to do it on a large scale. The Nile tilapia is one of the tropical fishes that can live in shallow water. It eats phytoplankton, periphyton, aquatic plants, tiny invertebrates, benthic animals, detritus, and bacterial films associated with detritus. Ponds achieve

sexual maturity at the age of 5-6 months (Chakraborty, Mazumdar, & Banerjee, 2010). When the water temperature hits 24°C, spawning begins. After the eggs hatch, the female incubates the eggs in her mouth and broods the fry until the yolk sac is absorbed. Fertility is linked to the female's body weight. A female weighing 100 gm will lay about 100 eggs per spawn, while one weighing 600-1000 gm will lay 1,000 to 1,500 eggs. Nile tilapia can live for more than ten years and weigh more than five kg. The biggest benefit of ponds is that fish can be bred easily. Tilapia culture is carried out in a variety of ponds. Low input ponds with unregulated breeding and sporadic harvesting are the most common but least productive; outputs are typically 500-2000 kg/ha/yr of unevenly proportioned fish. If monosex fish are stocked and consistent manuring and extra feeding are done, yields of even-sized fish can reach 8,000 kg/ha/year (Rathore et al., 2021). Agriculture and animal farming are both heavily involved in tilapia polyculture with other local fishes in freshwater ponds.

6.3 MARKET

In recent years, India's tilapia production has increased. The introduced improved tilapia strains are being supported in pond, tank, and cage cultures to compete in the domestic and worldwide markets. Modern techniques such as cages and biofloc technology are being used to increase culture productivity. Simultaneously, a number of challenges like unregulated culture and breeding techniques, unethical culture technologies, and poor attitude to disease management are surfacing, all of which require rapid attention in order to ensure the long-term viability of tilapia intensive production. Proposals for future tilapia farming development are made after considering the promises and prospects of intensive tilapia culture. Responsible stock production and distribution, better designed floating feeds, deployment of sophisticated aquaculture production methods, monosex culture, continuous quarantine and tighter biosecurity standards are all required. Early tilapia farming relied on a variety of production techniques, including fertilization, low stocking numbers, and supplemental feeding. However, the Mozambique tilapia or *Oreochromis mossambicus* that was first introduced failed due to early maturation, a lack of control over mating, and a poor growth rate that resulted in decreased production. Larger-scale research and development attention began in the late 1990s when tilapia farming's early success was mostly due to technological advancements that enabled aquaculturists to produce significantly higher yields. Controlling unwanted reproduction was also a problem, despite the fact that manual culling of females was initially the most straightforward technique to ensure the selection of faster-growing male tilapias. This procedure, however, was time-consuming and prone to errors.

6.4 CULTURE

O. niloticus (Figure 6.2) culture is presently in full swing, especially in Tamil Nadu, Kerala, Andhra Pradesh, Orissa, Maharashtra, Rajasthan, and West Bengal. Semi-intensive and intense production technologies have superseded traditional extensive cultivation methods. Tilapia are very appealing as a co-cultured fish because of their ability to improve water quality by feeding on plankton and detritus and modifying harmful bacterial populations while improving marketable production, especially in penaeid shrimp ponds. The culture of GIFT was successful in the biofloc system, since BFT aided in the fish's optimal growth and immunological state. The use of a biofloc-based zero water exchange technology to assure biosecurity and environmental sustainability in agricultural and inland saline water farming has been tried and proven. The GIFT strain of Tilapia has been deemed the optimum stocking density at higher stocking densities of 200–250 m^3, however higher stocking densities of up to 350/m^3 can also be grown in biofloc systems without compromising growth and immunity(Elangovan Prabu, Felix, & Uma, 2021).

Figure 6.2: Mono sex tilapia.

Source: https://nas.er.usgs.gov/XIMAGESERVERX/2012/20120530095223.jpg

Despite the fact that tilapia is typically produced as a monoculture, tilapia aquaculture has also been embraced as a polyculture with appropriate indigenous fish species in order to continue increasing production and improving sustainability. Tilapia has also been used in shrimp polyculture, offering an extra revenue crop while simultaneously lowering the prevalence and severity of viral and bacterial diseases in shrimp. Farmers have introduced tilapia as an extra polyculture species to traditional carp ponds in pond culture. In the states of Andhra Pradesh, Gujarat, and Tamil Nadu, tilapia polyculture with carps is possible. Polyculture of shrimp and carp accounts for 3-5 percent of national tilapia production (Lalramchhani et al., 2019). Tilapia was the first species widely introduced for aquaculture, whereas O. niloticus was introduced into India in late 1987. The aquaculture of O. niloticus spread across the country, owing to the efforts of private entrepreneurs. The yield of golden tilapia was estimated to be 10 tonnes every six-month culture cycle. Production, on the other hand, came to a halt for unclear causes or due to a lack of consistent seed supply. Red tilapia production and sales increased as a result of growing acceptance of red types and acknowledgment of the value they add to the finished product.

Figure 6.3: Cage culture.

Source: https://fisheries.tamu.edu/files/2021/02/cage-culture.jpg

GIFT, on the other hand, is still a popular tilapia on the market. The WorldFish Center in Penang, Malaysia, established the Genetically Improved Farmed Tilapia (GIFT) culture program, which has now been passed on to private business for further enhancement of the fish in India. MPEDA has introduced GIFT to enhance exports, with plans to generate 12 lakh metric tonnes of GIFT worth 8000 crores INR in the next few years. Tilapia fry are typically grown in nursery ponds until they reach a weight of 20-40 g for 5 to 8 weeks before being refilled into grow-out ponds (Srinivas, 2022).

Figure 6.4: Pen culture.

Source: https://www.fao.org/3/ad021e/AD021E25.jpg

In India, the introduction of cage (Figure 6.3) and pen (Figure 6.4) cultures has enhanced reservoir fish production significantly. In areas with suitable water quality, flushing rates, and water depth, floating cages have become the most popular system for rearing tilapia in India after *Pangasianodon hypophthalmus*. The high nutrient inputs that a water body receives in the form of wasted feed and metabolic wastes may induce eutrophication and unintended releases of fish that are prone to causing ecosystem damage. Before such actions are disseminated, intensified, and consolidated, it is critical to do a scientific investigation in order to build a safe technique. At the same time, physical impediments to traditional fishermen's fishing activities in these water bodies, as well as the resulting conflicts caused by tilapia infestations, should be mitigated through studies. State fisheries agencies have been using leasing and licensing procedures to allow cage culture in public waters without a scientific foundation or taking social concerns into account. Thousands of fish farmers have built bheris, or fishponds, across 12,500 hectares of sewage water on the eastern outskirts of Kolkata, West Bengal. Tilapia yields from wetlands/bheries average around 750 kg per hectare per year, with individual growth rates of 150-200 g in 3-4 months. Around 50,000 growers and fish traders are supported by existing bheris or fishponds. The majority of them are small-time private business owners that make a living by raising 10,000 tonnes of wastewater-fed fish every year(Bainy, Bertan, Corazza, & Lenzi, 2015).

The development of monosex male tilapia was one of the most important commercial innovations in the early 2000s. To establish all-male populations, Tilapia fingerlings are administered the androgen 17--methyltestosterone in their diet. The spread of sex-reversal technology led to an increase in tilapia output all over the world. Privately held tilapia hatcheries are the most common, and stand-alone enterprises sell to a network of grow-out farms. Controlling unwanted reproduction was also a problem, despite the fact that culling females by hand was initially the simplest way to ensure the selection of faster-growing male tilapias. This procedure, however, was time-consuming and prone to errors. Following that, hybrid production technologies gave tilapias a boost in terms of early sexual development and prolific reproduction. In the early 1990s, the introduction of the faster-growing and slow maturing Nile tilapia resulted in the formation of various inter-specific hybrids, including golden tilapia. Captive hybridization and quick technology distribution resulted in more tilapia hybrids becoming available for aquaculture later.

There are five species of tilapia in India; the Oreochromis mossambicus was the first tilapia species to be imported in 1952. It immediately earned the reputation of being a pest due to its prolific breeding. Later, O. niloticus, also known as Nile tilapia, became the preferred species for culture (E. Prabu, Rajagopalsamy, Ahilan, Jeevagan, & Renuhadevi, 2019). In 2009, the Indian government approved aquaculture, stipulating specific rules as part of a species diversification program aimed at raising the country's overall fish output levels. Because the fish is at the bottom of the food chain, its culture is considered cost-effective and environmentally friendly. According to current criteria, only monosex male/sterile via hormonal manipulation or cross-breeding fish can be farmed, and the species recommended are Nile tilapia and, more recently, genetically enhanced farmed (GIFT) fish. The third species found in India is O. urolepis, (Figure 6.5) which grows in ponds and tanks in Andaman and Nicobar and has a production capacity of over 1,150 kg/ha/culture cycle. The fourth species found in Chennai is O. aureus, which was discovered and identified by chance. The fifth species is Tilapia zilli, which can be found in vast numbers in the Indira Gandhi canal (Rajasthan) and is not farmed.

Figure 6.5: O. urolepis.

Source: https://nas.er.usgs.gov/XIMAGESERVERX/2005/20051102135129.jpg

6.5 PROGRESS

Although tilapia aquaculture is a relatively new endeavor, it has made tremendous progress in commercial fish farming in India, with the goal of meeting the enormous demand for the fish in both the domestic and international markets. In India, the introduction of cage and pen culture has enhanced reservoir fish production significantly. After *Pangasianodon hypophthalmus*, floating cages have become the most popular technology for producing tilapia in India in places with acceptable water quality, flushing rates, and water depth. China, Indonesia, Brazil, Bangladesh, and India are among the countries where this practice is becoming more popular. Because Indian reservoirs and lakes have significant potential for tilapia cultivation, MPEDA is working hard to increase tilapia output in cages.

With the technical assistance of MPEDA, private fish exporters aim to invest over INR 100 crore in cage-rearing tilapia in lakes in Maharashtra and Rajasthan over the next four years. Although tilapia culture in cages erected in reservoirs or lakes has the potential to increase productivity, it has environmental and social consequences. The excessive nutrient inputs that a water body receives in the form of wasted feed and metabolic wastes may cause major environmental problems such as eutrophication and inadvertent fish release, causing ecological damage.

The male tilapia is highly adapted to supplementary feeding and has a high profit margin in monosex tilapia commercial farming due to its quick growth. In both domestic and foreign markets, this fish is in high demand. To sex reverse fry for all-male production, existing operating hatcheries use 17-methyl-testosterone. The majority of tilapia hatcheries are small, privately owned businesses that sell to a network of grow-out farmers. Farmers in India are reliant on monosex production technology from outside the country, therefore a superior easy and generic method of monosex production of tilapia has just been described. Because tilapia is a resilient fish, disease occurrence is rare. Despite this, various bacterial diseases have been reported in tilapia culture, especially in indoor systems. Antibiotics, prebiotics, and probiotics are commonly used in tilapia culture, which is an increasing source of concern. The effects of antibiotics, probiotics, prebiotics, and alternative feed ingredients on the intestinal microbiota of tilapia in advanced tilapia culture techniques have been studied, and it is suggested that more research into the microbial ecology, alternative feedstuff effects, and economic impacts of modulating tilapia intestinal microbiota is needed to develop a sustainable tilapia production technology. Antibiotic resistance in bacteria isolated from tilapia farms is also a concern as zoonotic pathogens. Furthermore, tilapia lake viral disease or TiLVD has just been discovered as a novel virus disease that is developing as a possible danger to the global tilapia business. The disease was initially discovered in Israel's freshwater lakes in 2014, killing a large number of tilapias. The virus was then discovered in freshwater and brackish water in Ecuador, Colombia, Egypt, Thailand, and India. The impact of tilapia lake virus disease on wild stocks might be huge, both economically and in terms of biodiversity and ecological consequences. Two tilapia cell lines have recently been produced for TiLV strategic control and prevention.

At the moment, aquaculturists favor tilapia culture since basic information and culture techniques are readily available, and they are simple to implement in order to compete in worldwide markets. Tilapia can be produced twice a year and is available even in some highland states. If current farming methods and technology are employed in tilapia fish farming, more income will be generated, providing a promising opportunity to contribute to our economy. Tilapia is the 2nd most farmed fish in the world, and due to the ease of aquaculture, marketability, and steady market pricing, output has doubled in the last decade. Nonetheless, refining of current technology is required to provide a robust genetic base, ultimate product quality, and effective farm management and health management in accordance with government rules and regulations. Tilapia production from pond, tank, and cage cultures has been greatly promoted, elevating commercial productions, thanks to the country's existing regulatory framework, the introduction of improved strains, and the application of quarantine and biosecurity. Product standards, technical standards, hygiene standards, and other applicable laws are still key challenges in tilapia farming. Excessive use of antibiotics and other chemical agents endangers aquaculture's long-term viability by compromising food safety and badly impacting environmental conditions. In recent years, voluntary sustainability standards and certification schemes have been developed to promote value chain transparency and allow customers to factor sustainability into their purchasing and consumption decisions. For example, requirements include the preservation of biodiversity, natural ecosystems, and water resources, the responsible use of aquaculture-related resources, the protection of fish and human health, and socially acceptable working conditions.

6.6 DISTRIBUTION

The distribution of genetically proven improved GIFT stocks, better formulated and floating feeds, deployment of advanced aquaculture production systems, continued quarantine and tighter biosecurity standards, as well as stringent monitoring, are all suggested as ways to improve tilapia production in India. Furthermore, using a proven monosex culture technique will increase the yield of all male tilapia cultures. There is also a critical requirement for technology refinement to assure stock enhancement, product quality, and appropriate management of tilapia-health hazards. Fish farmers and other aquaculture stakeholders will benefit from advances in information and communication technology as they embrace improved management techniques. These advancements should include high-tech modeling and monitoring tools for improving tilapia farming's environmental and economic performance. To ensure the sector's long-term viability, the relevant regulatory measures must be properly implemented. If necessary, the existing guideline should be reviewed for changes and amendments, as well as stronger enforcement. The introduction of enhanced tilapia strains into pond, tank, and cage culture has resulted in increased aquaculture production in India. Today, tilapia farming with cages and biofloc technology is popular, and aquaculturists are attempting to break into the international market as well as meet domestic demand. Looking at the future of tilapia culture in India, distribution of improved stocks, better formulated and floating feeds, deployment of advanced aquaculture production systems, monosex culture, and continued quarantine and tighter biosecurity standards are suggested as action points for further improvement of tilapia production in India.

6.7 CAGE AQUACULTURE

Cage aquaculture is the practice of raising fish in enclosed structures that allow for free water exchange. Even though fish breeding in cages has been practiced in Southeast Asian countries since the late 1800s, the technology has only recently acquired appeal among Indian fish growers and fishermen. The Central Marine Fisheries Research Institute (CMFRI) was the first in India to establish open sea cage farming and popularize it through frontline demonstrations in several maritime states. When compared to pond fish culture, cage fish farming is a low-impact agricultural approach with excellent economic returns, providing 10-12 times the yields(Jayasankar, 2018). Cage culture was established in estuarine and coastal areas throughout the country after successful demonstrations in the sea. Cage farming has enormous potential for increasing fish output in the country's marine, estuarine, and coastal waterways. Large-scale cage farming thrives in the coastal waters because of the abundant water resources and favorable environmental conditions. Multispecies cage farming can be commercialized in the nation with minimal expenditure. Cage farming is a highly profitable alternative occupation that fishermen can pursue individually or in groups. Cage aquaculture is a widely accepted farming approach due to its low initial investment cost and recurring expenditure, ease of operation, short culture time, and excellent returns per unit volume.

The economic viability of any farming technique is critical to its acceptance. When compared to shore-based farming approaches, the initial investment and recurring expense related to infrastructure creation and maintenance are cheap in cage farming. Cage farming generates a high level of gross income since it cultivates high-value fish that are in high demand both domestically and internationally. Due to its excellent quality and freshness, cage-farmed fishes have a wide range of market potential in both the local and export sectors. Farmers and fisherfolk can use economic viability and financial feasibility research to make investment decisions and use scarce resources efficiently to maximize economic advantages. Economic data can also help policymakers and financial institutions design lending rules, repayment choices, and sector subsidies. In the sea, estuaries, and coastal waterways, the CMFRI has defined cage farming techniques for several types of cages.

The choice of a good site for cage farming is a critical factor in determining the economic success of cage farming. It determines cage farming's success, as well as capital investment, operating costs, yield, and mortality rates. The installation of cages should be avoided in areas where aquaculture activities are forbidden by regulatory authorities. Other uses, such as nautical routes, fisheries, tourism, or marine protected areas, should be avoided. Prior to the installation of cages, attention must be taken to guarantee that fisherman organizations, self-help groups, or fishermen cooperatives have proper leasing rights or legally acceptable authorization. Before choosing cage sites, environmental or topographical factors should be taken into consideration. The depth of the water body, carrying capacity, water quality requirements, winds, waves, currents, tides, and bottom substrate are all important considerations to consider while choosing a cage site. Geographical information systems can be used to analyze environmental challenges in coastal zones and determine if they are suitable for aquaculture. Strong winds created by cyclones can destroy cage structures as well as fish placed in cages, however moderate wind is favorable to cage farms. Cages should not be installed in areas where there is a lot of wind. Winds should not exceed 10 knots for floating cages in the sea. Wind velocity affects wave height, and wave energy rises proportionately with the square of wave height. The highest wave height is 1 meter. A mild, continuous current stream is beneficial to cage farming, whereas excessive currents damage floating structures, cages, and has a negative impact on fish behavior. For cage farming, the maximum current velocity allowed is 0.05-1 ms^{-1}. Many sites around the Indian coast have current speeds of up to 1.2 ms^{-1}, and such sites should be avoided for marine cage farming. Similarly, for marine cage culture, a tide amplitude of 1m is ideal(G. S. Rao, Imelda,

Philipose, & Mojjada, 2013). The monsoon season is avoided for marine cage farming due to the unpredictability of current velocity during this time.

Figure 6.6: Cage production system.

Source: https://www.researchgate.net/profile/Patrick-Trail/publication/340095336/figure/fig5/AS:872115070857216@1584939844067/Newly-established-cage-culture-systems-for-the-production-of-Tilapia_W640.jpg

The cage production system (Figure 6.6) consists of a floating structure, net materials, and mooring system with a round or square shaped net cage to retain and raise fish. It can be deployed in the sea, lakes, rivers, or reservoirs. Capital and operational expenditures are the two key cost components in cage farming. The cage structure, nets, accessories, mooring, and installation expenses are all included in the capital cost component. Miscellaneous items for storage and transportation of feed and fish are included in the capital cost category. Depreciation and interest on fixed capital are subtracted from the total investment cost to arrive at an annual fixed cost. Cage frame depreciation is computed based on a 7-year expected life for sea cages and 5-years for backwater and estuarine cages. The depreciation on nets, floats, and accessories was computed based on a five-year projected life. Feed, seed, labor charges for feeding and harvesting, and cage frame and accessory maintenance fees are among the primary operational cost components. The expenses and revenues in open sea, estuarine, and brackish water regions differed in terms of cage frame size and materials, accessories and mooring system, types of fish cultured, stocking density, feed, and maintenance expenditures.

Feed, seed, labor expenditures, boat rental, harvesting, cage frame and accessory maintenance are the major components of operational costs. Sea cage farming is suitable for Cobia or *Rachycentron canadum*, Silver Pompano or *Trachinotus blochii*, Asian Seabass or *Lates calcarifer*, Snappers or *Lutjanus* sp., Groupers or *Epinephelus* sp. and Spiny Lobster or *Panulirus* sp. The price of seeds varies depending on the size of the fingerlings and the density of the stocking. Depending on the size of the fingerlings, the typical cost per seed for cobia, pompano, and seabass ranges from $20 to $50. A few government-owned and private hatcheries in the country produce cobia, pompano, Asian seabass, and grouper seeds. Like any other aquaculture business, feed and feeding account for a considerable portion of cage farming's operational costs. Feed costs account for roughly 50-75 percent of total operating costs. The economic benefits of cage farming can be maximized by producing cost-effective feeds and intelligent feeding schedules. It also reduces feed waste and contamination in the environment. For juveniles, a feeding rate of 10% of body weight is advised, which can be reduced to 3% as the fish grows and progresses in culture. Feed costs range from $20-25 per kg for garbage fish to $75-90 per kg for designed feed(Sumithra et al., 2019). Feeds must be stored properly in order to keep their quality. For feed storage, cold storage preservation or freezers are preferred to maintain quality. Cage maintenance entails routine monitoring for food adjustments, environmental parameter monitoring, disease monitoring, and predator monitoring in order to reduce operational costs and maximize economic gains. The entire construction, including the cage frame and mooring, should be inspected on a regular basis, and essential maintenance and repairs should be performed. To ensure water quality and allow faster growth of fish, timely replacement of net cages, cleaning of nets, and mending of damaged nets are also required.

Cage farming has a lot of potential for increasing fish production in the country's coastal and brackish water zones. The states of Kerala, Karnataka, Goa, Maharashtra, Odisha, and West Bengal have 1.2 million hectares of brackish water areas.

Figure 6.7: Asian seabass.

Source:https://upload.wikimedia.org/wikipedia/commons/5/50/Lates_calcarifer%2C_2014-09-19a.jpg

Figure 6.8: Etroplus suratensis.

Source: https://www.fishbase.de/images/species/Etsur_u4.jpg

Asian seabass (Figure 6.7) or *Lates calcarifer*, pearl spot or *Etroplus suratensis*, (Figure 6.8) tilapia or *Oreochromis* sp, mullet or *Mugil cephalus*, red snapper, and caranx are the most common species appropriate for brackish water production. Mullet and Tilapia can be raised in cages with minimal seed and feed costs. The country lacks seed production techniques for mullet, red snapper or *Lutjanus argentimaculatus*, and *Caranx* sp, hence these fish are cultivated using catch-based aquaculture, in which juveniles are collected alive and allowed to grow to marketable size in cages. Farmers used a variety of cage sizes for farming in coastal waterways, including 2x2x1.5m^3, 4x4x2m^3, 8x4x2m^3, 4x4x4m^3, 8x4x4m^3, and 6x6x4m^3(M. Rao et al., 2021). In coastal waters, a cage size of 48m^3 is recommended for operational efficiency and profitability. The stocking density in coastal locations varies depending on cage volume and fish type. In coastal waters, composite seabass cultivation with pearl spot is favored since the latter aids in net cleaning and offers farmers with better market options and profits. For the purpose of comparing profitability, the economic performance of cage farming of various fish species in various cage dimensions was estimated. It was discovered that growing a composite of seabass and pearl spot was more profitable than farming a single species. Pearl spot develops faster in cages, has highly nice flesh, and has a lot of market potential.

In India, the estuary waters are quite fertile and serve as a nursery for a variety of fish. Climate change has resulted in the salinization of estuaries and a decrease in fish catches, posing a threat to small-scale fishermen that rely on estuarine fisheries for a living. Cage fish farming in estuarine waters can be advocated as a viable option for small-scale fishermen that rely on estuarine fisheries for a living. In the estuaries, seabass, snappers, pearl spot, and carangids can be raised. The Central Marine Fisheries Research Institute has stepped forward to promote cage fish farming in the country's estuary waterways. Nylon ropes are typically utilized as mooring ropes, with sandbags used to anchor the cages(Mondal, Shahin, Wahab, Asaduzzaman, & Yang, 2010). The stocking density is 50 fish per m^3 and the fish are typically 10 months old. Efficient marketing channels are critical components of farming activities that are economically viable. The country's expanding demand for high-quality fish products, combined with diminishing catches from marine capture fisheries, presents huge prospects for farmed fish marketing. Due to their higher quality and freshness, cage farmed fish are generally sold through local fish markets or at farm gates, and command a premium price. Various institutional organizations active in promoting cage farming in the country, including the CMFRI, state fisheries departments, cooperative banks, and non-governmental organizations, engage in market promotion efforts such as online portals, live fish sales, and fish harvest melas. However, large-scale cage farming expansion involves investigating improved marketing options in both domestic and international markets. For addressing future marketing challenges, capacity building of small-scale fishermen or fish farmers in the country to improve entrepreneurial capabilities, market promotion through fishermen/farmers co-operatives or Farmer Producer Companies (FPOs), better storage and transportation infrastructure, and value-added products, among other things, should be promoted.

6.8 RESEARCH

Cage fish farming offers enormous potential for providing revenue and employment to the coastal fishing community. With CMFRI's technical assistance, more than 3000 cages have been installed across the country's maritime states. These cage farms could produce 5,250 tonnes of fish, giving a total net benefit of 105 crores for cage fish farmers(Aswathy, Imelda, Ignatius, & Joseph, 2020). Cage farming will also benefit the input suppliers and sellers of cage frames and accessories. Cage farming is promoted with financial aid from the federal government and state fisheries agencies. Through different initiatives, the Karnataka government also gives financial help to promote marine and brackish water cage farming. The growing interest in extending priority sector funding for cage farming by the banking sector in the country, notably the National Bank for Agriculture and Rural Development (NABARD), demonstrates the sector's future prospects. Cage fish farming is a lucrative business in the country's marine, estuarine, and brackish water zones. Large unused regions in the sea, estuaries, and brackish water areas offer great opportunities for increasing fish production in the country through cage farming. However, a key stumbling block to large-scale cage fish farming in the open sea is a lack of leasing rules and regulatory measures. As a result, creating policies and regulatory measures that take environmental carrying capacity and socio-ecological aspects into account is critical. For large-scale commercialization of cage farming in the open sea and coastal waters, insurance plans for minimizing risks due to natural calamities or anthropogenic activities are also required.

CHAPTER 7

FARMING OF TILAPIA IN SRI LANKA

CONTENTS

7.1 INTRODUCTION

During the years 2011-2014, the Government of Sri Lanka, through the Ministry of Economic Development, undertook an island-wide multi-sector program including rural households with the goal of improving nutrition as well as providing extra income and livelihoods (Bandara et al., 2008). This was the government's second attempt to implement small-scale pond fish farming in Sri Lanka, after a failed initial attempt in 1980-1983. The current initiative has resulted in the establishment of 2,109 ponds spanning a total area of 105.45 ha. Through the use of a questionnaire, information on cultural techniques, fish productivity, financial returns, and costs associated with these ponds were gathered. The only species employed for aquaculture was the Nile Tilapia or Oreochrimis niloticus. In these ponds, all-male or mixed-sex tilapia are raised. There were significant differences in stocking densities, inputs, fish output, fish selling prices, operational costs, and profitability among ponds in these districts. The average fish production per culture cycle in Gampaha was 9,192 kg/ha, 6,074 kg/ha in Puttalam, and 3,270 kg/ha in Puttalam (Ampara) (Pet, Gevers, Van Densen, & Vijverberg, 1996). In all three districts, pond tilapia farming is profitable. In the case of ponds in Gampaha and Puttalam, profit margins are particularly good. (Figure 7.1)

Figure 7.1: Tilapia aquaculture in Srilanka.

Source:https://www.sevalanka.org/uploads/6/6/1/2/66123167/dscf7554_1_orig.jpg

7.2 FISH POND FARMING

Fish pond farming (Figure 7.2) is a centuries-old tradition in many nations throughout the world, particularly in Asia. Nonetheless, there is no tradition of fish farming in Sri Lanka. During the years 1980-1983, Sri Lanka's Ministry of Fisheries attempted to popularize freshwater fish farming by introducing small-scale pond fish farming. Subsidies in the form of financial grants were provided as part of this initiative to cover a portion of the cost of the ponds and aquaculture installations. Although there was a lot of interest in fish farming as a result of this subsidy plan, and roughly 1500 pond units were built, practically all of these ponds were abandoned within a few years (Munasinghe, Stephen, Abeynayake, & Abeygunawardena, 2010).

Figure 7.2: Fish Pond farming.

Source:https://exploresrilanka.lk/wp-content/uploads/2020/09/2-e1600237079953.jpg

During 2011-2014, the Government of Sri Lanka, through the Ministry of Economic Development, undertook an island-wide multi-sector Divi Neguma program including rural households with the goal of improving nutrition as well as generating additional income and livelihoods. One of the program's components was home-garden pond fish farming, which was overseen by the National Aquaculture Development Authority of Sri Lanka (NAQDA), which is part of the Ministry of Fisheries and Aquatic Resources

Development. People were encouraged to start small-scale fish farming in ponds by the government, which provided financial aid to cover the cost of pond construction. In addition, potential fish farmers were given hands-on training in fish farming and were supported by the NAQDA's extension department.

A total of 2,109 ponds covering an area of 105.45 ha have been constructed as part of this operation. Aquaculture is said to be a centuries-old practice that began in China. However, it is just three to four decades old that it has made a significant contribution to human nourishment. Aquaculture generated 66.2 million tonnes of edible fish in 2012, out of a total of 136.2 million tonnes accessible for human consumption around the world. In 2012, the Asia-Pacific region led the globe in aquaculture production, accounting for more than 89 percent of total worldwide farmed fish production(Thilakaratne, Rajapaksha, Hewakopara, Rajapakse, & Faizal, 2003). Freshwater fin fish account for almost two-thirds of Asia and the Pacific's farmed food fish production. Many Asian and Pacific countries rely on freshwater omnivore and herbivorous fish as a source of protein. Pond culture is still the most common method of producing fin fish, accounting for more than 80% of global aquaculture production. Despite the fact that pond fish farming is very popular in several Asian nations, Sri Lanka remains far behind in this regard. Attempts by the Sri Lankan government to establish small-scale fish farming in ponds between 1980 and 1985 were a failure. Fisheries officials blamed the failure on a lack of fish farming history, a lack of fish farming knowledge transfer to farmers, a lack of extension service, and low financial returns from small pond culture systems due to low yield and low prices for freshwater fish species at the time. After roughly 30 years, the second effort at introducing food fish farming in ponds was made. With a total coastline of around 1,700 kilometers, Sri Lanka is ideally suited for aquaculture. Lagoons and estuaries cover a total area of 121,000 hectares. A large tract of low-lying delta lands, estimated at 70,000 hectares, borders these estuaries and lagoons(Ukenye, Taiwo, & Anyanwu, 2019).

In Sri Lanka, the aquaculture sector began in the early 1980s, when a few large multinational businesses and a few small-scale entrepreneurs began producing shrimp. The pioneer phase progressed slowly until the early 1990s due to political turmoil. However, because of favorable public investment incentives and substantial economic returns, the shrimp farming sector grew rapidly and uncontrollably between 1992 and 1996. During this time, they produced 8,000–9,000 kg/ha/year in traditional earthen ponds(Hirimuthugoda, Liyanage, & Amaraweera, 2014). Small-scale farms of 0.5–1.0 hectares were built in clusters, mostly near lagoons and environmentally sensitive areas like mangroves, salt marshes, inter-tidal mud flats, coconut plantations. However, due to a lack of appropriate law enforcement and poor environmental planning, the rapid development resulted in about 47 percent of all farms being unlicensed. Sri Lanka currently has a low but consistent shrimp production and an increasing fin fish aquaculture production. The main focus has been on starting small, with fish farming attempting to provide revenue and food to small local populations. Several trial projects, including sea cages and land-based installations for fish and shrimp, have been developed in various parts of the country in recent years. Based on present cage farming expertise, there is significant interest in erecting larger marine cages with circumferences of up to 60 meters in the island's northeast. Because Sri Lanka is only 7 degrees north of the equator, its seas are calmer and less vulnerable to natural calamities like typhoons, storms, etc. than any other Asian countries. Because maximum wave heights range from 3.5 to 4.0 meters, sea cage farming has a lot of potential. In Sri Lanka, the land-based farming industry is also getting more attention. Farmers are currently following a stringent coastal zone management plan limiting the period for stocking and harvesting in different farming locations, and traditional earthen pond farming of shrimp and finfish is progressively rising. With good environmental monitoring and control, stocking densities are steadily increasing. In addition, numerous international technology providers, especially Norwegian firms, have recently launched and constructed larger commercial production units for Barramundi and Tilapia in recirculating aquaculture systems (RAS) (Harkes, Drengstig, Kumara, Jayasinghe, & Huxham, 2015).

7.3 ADVANCEMENTS

The level of regularly used technology is also undergoing significant advancements at the present, including the introduction of biofloc systems. During this decade, mitigation strategies such as real-time environmental alerts have been tested out. Shrimp farmers in Chilaw, the North-Western Province's primary shrimp-farming center, have successfully tested a novel remote alert solution to avoid or minimize negative effects on production from nearby areas/farmers. The National Aquaculture Development Authority (NAQDA) found that about 8,500 hectares of land in Sri Lanka might be utilized for aquaculture(Lakpawan, Liyanage, Tennakoon, Herath, & Kulathilaka, 2019). Only about a quarter of these areas are now exploited for aquaculture of diverse species, despite the fact that feasibility studies have shown that aquaculture may be done in these places with less biological risks. The Northern Province's water quality was also assessed by NAQDA, which found appropriate levels of dissolved oxygen, pH, salinity, and alkalinity. In the fishing sector, Sri Lanka has created a robust infrastructure that is appropriate for large-scale export to markets prepared to pay for high-quality seafood. At the moment, Sri Lanka is becoming a popular destination for international aquaculture investment. Norway, Scotland, Vietnam, Japan, the United States, and Canada have all invested in the last ten years. Hatcheries, land-based on-growing systems, sea cages, and the building of infrastructural facilities such as roads, common inlet/outlet canals for water intake and release, sedimentation canals, ponds, electricity, and so on are all receiving funding. Furthermore, with the help of the Sri Lankan government, links between the government, breeders, farm-zone managers, and academic experts have been developed. This enables effective information sharing between farmers, with input from farmers providing relevant and practical information to the Ministry of Fisheries, which is used to translate generic Best-Management-Practices into local policies, plans, and actions.

Sri Lanka intends to improve its aquaculture value chain, which includes broodstock and domestication, juvenile production, feed development and production, on-growing, and commercialization. Sri Lanka aspires to become a key player in the region's burgeoning aquaculture industry, because of its unique geographical location in South-East Asia. The recent development in the northern parts of the island is creating new opportunities by allowing access to many regions that are ideal for brackish and marine aquaculture. For foreign aquaculture technology suppliers, investors, and distributors, this is a good commercial opportunity. Aquaculture is transitioning from a subsector to a concentrated commodity production, striving for superior technology and long-term development in all water bodies. The future potential for aquaculture in Sri Lanka is, of course, unique in Asia, as it has the ability to cut wild-caught supplies from 90% to 10%, with aquaculture filling the gap. In addition, the *L. vannamei* high-yield prawn strain, commonly known as pacific white or white leg shrimp, was introduced to Sri Lanka in 2018 and is now in increased demand worldwide. Imports of *L. vannamei* account for about 85% of total volumes in the European Union, with the remainder coming from black tiger or *P. monodon*. Simultaneously, the industry is advised to adopt Critical Better Management Practices or CBMPs, which include, among other things, zero water exchange or Biofloc technology.

Many Norwegian projects have been launched in the last 20 years to transfer Norwegian technologies and know-how to Sri Lanka. For a stable scale-up to industrial aquaculture, Sri Lanka will need to build a complete and well-functioning value chain in the coming years(Bentsen et al., 2017). The government of Sri Lanka sees aquaculture expansion as a way to promote rural development and economic stabilization, both of which are important for food security. With the end of the civil war, new opportunities for aquaculture development have arisen. Aquaculture benefits refugees from war-torn places as well as the rural poor. Our mission was to assist in the identification of strategic goals for sustainable aquaculture growth and the development of knowledge-based capacity for farm-level practices as well as local, regional, and national policies. The main hurdles to the application of practices and policies were a lack of information and poor knowledge connections, rather than a lack of resources or technology. Scoping and systematic literature reviews; policy review; farmer and farm household surveys; on-farm observational studies; multi-criteria decision analysis; pilot projects in new aquaculture methods for smallholder farmers; application of models for selecting production sites; network analysis; and knowledge mobilization interventions were among the approaches and methods used.

Figure 7.3: Shrimp aquaculture.

Source:https://62e528761d0685343e1c-f3d1b99a743ffa4142d9d7f1978d9686.ssl.cf2.rackcdn.com/files/89033/area14mp/image-20150720-12558-11zor9d.jpg

7.4 SHRIMP

Shrimp aquaculture, (Figure 7.3) for the most part, improved food security by providing revenue rather than directly adding to household food consumption. Working with communities in the NWP to improve fish production from seasonal reservoirs addressed challenges in maximizing fingerling stocking and demonstrated that local fish availability, as well as providing broader local markets, may contribute to food security and rural income. The oyster production pilot project in Puttalum Lagoon with a community of 35 families showed substantial potential for generating local revenue for coastal communities facing dwindling capture fisheries resources. The project's experience with lobster fishing communities revealed a significant economic potential for a change toward the fattening of undersized animals and eventually full culture. Small-scale ornamental fish farming has also flourished in Sri Lanka, contributing significantly to rural income. As a result, aquaculture in Sri Lanka is showing promise for rural development. The project's most important finding was the importance of fish and shrimp infections as a persistent source of output instability and farm failure. This was looked at the most in the shrimp sector, and the findings influenced a lot of the subsequent project work. In Sri Lanka, all sectors of aquaculture suffered from a lack of capacity to assess, diagnose, manage, and prevent infections. The inability to control disease and apply good procedures make shrimp producers feel powerless over their farm's fate, resulting in severe farm-level production instability. As a result, initiatives in shrimp farming must prioritize this issue.

Knowledge sources and skills were poorly connected, restricting the dissemination of knowledge about sustainable production techniques, structures, and strategies to farmers. The EP, which is a new aquaculture development region, was split from the NWP, which has a longer history of aquaculture. Farmers primarily relied on neighbors and other farmers for information, rather than government agencies or institutions. Few farmers and industry people had formal aquaculture education, which was a significant issue because farmers with formal education reported more sustainable practices. Farmers with broader social networks may have reported more sustainable practices, according to ongoing research. Farmers were unable to embrace better management practices (BMPs) designed to assist them to survive a crop cycle due to a lack of trustworthy relationships and market pressures. Some of the hurdles to knowledge sharing were resolved through project activities that included participatory BMP creation and better knowledge connectedness through the distribution of timely information to farmers via mobile SMS. Farmers' knowledge and self-efficacy to modify practices and boost cooperative actions in response to emergent challenges increased in a pilot SMS intervention with 60 farmers. The NWP and EP are adapting these ideas for use in other aquaculture sectors, especially CBF, as well as emerging potential like oyster farming.

7.5 SUSTAINABLE AQUACULTURE

Sustainable aquaculture (Figure 7.4) strategies were either non-existent or scant in Sri Lankan legislation, despite their support in principle. There were no strategies in place for policy implementation, monitoring, or enforcement. The shrimp industry has evolved under a constant attack of disease, resulting in value chain organizational structures that are unstable.

Figure 7.4: Sustainable aquaculture.

Source: https://www.sundayobserver.lk/sites/default/files/styles/large/public/news/2018/10/19/z_p45-Sustainable.jpg?itok=Ht7y_gme

There were no strategies in place to adequately engage women in aquaculture. The small number of female-run farms discovered in the surveys reflected this. The policy evaluation identified important factors to consider when evaluating policies and produced a gap analysis that was shared with the government. The purpose of the foundational research was to establish and evaluate a set of indicators for sustainable shrimp farming that can be used by industry and government to determine policy priorities and track progress toward on-farm sustainability targets. It was discovered that participatory policy formulation is not the norm. This resulted in issues with disease BMP uptake due to a lack of consideration for local limits in applying BMPs and a lack of ownership of BMPs by farmers. To investigate a significant bottleneck for culture-based fisheries, a multi-stakeholder participatory process was done (CBF). The approach was successful in finding an accepted set of development alternatives for this sector in the EP, and the NWP administration intends to replicate it in the NWP.

Aquaculture has the potential to be a reliable source of consistent income and, as a result, an underlying component of Sri Lanka's rural food security. However, that promise will not be realized until there is an enabling policy environment, emphasizing co-management, developing specific plans and actions aimed at fostering sustainable aquaculture; the threat of disease is addressed; and knowledge is better connected and mobilized along the value chain with government and university. Because aquaculture is growing in those sectors, the lessons learned from building the industry in the NWP must be more effectively synthesized and disseminated to the EP and Northern Province, or the risk of repeating the NWP's shortcomings and mistakes is extremely significant. The tilapia *O. mossambicus* and its hybrid *O. niloticus* are estimated to account for 70-95 percent of the total inland fisheries yield, providing a significant source of low-cost, high-value food for the rural poor. Because both marine and inland fisheries in major reservoirs are thought to have achieved their maximum sustainable output, innovative approaches for enhancing inland water productivity without jeopardizing the existing tilapia fishery are urgently needed. There are two different tactics that can be used. The first is based on exploiting a large biomass of previously unexploited indigenous minor cyprinids, while the second is centered on improving fisheries through culture-based stocking programs. Minor cyprinids have an output potential of 250-300 kg/ha/year, which is comparable to the productivity of the commercial tilapia fishery(Amarasinghe, 1990). According to current research, a number of marketing obstacles are expected to be a key deterrent to this pathway's adoption.

7.6 YIELDS

Tilapia and culture-based carp yields have been found to be positively and adversely linked with reservoir area, respectively. Furthermore, it has been claimed that the carp fishery has a deleterious impact on the tilapia fishery. Other trials have shown that culture-based fisheries, even in smaller perennial tanks, may not be feasible where established artisanal tilapia fisheries already exist. The smallest seasonal tanks, which are highly controllable and display high natural output, have shown the greatest technical promise. Many of the smallest of these tanks are accessible to those of modest income. Although they are high-potential systems, the natural output is limited when recruitment is disrupted on a recurrent basis during dry years. In areas where aquaculture is not practiced, low-risk enhanced fisheries based on seasonal stocking programs may be justified. In contrast to existing government policy, which prioritizes stocking hatchery seed in bigger systems, it is recommended that hatchery seed be favored for such seasonal tanks.

Poor seed availability continues to stop the efforts to revitalize the program. There is currently no private sector production or market, and most of the state's production facilities have long been leased to a more profitable decorative export industry. This potential cost shows that, under current conditions, private sector adoption of food fish seed production is unlikely. Community fingerling production programs, which will produce advanced fingerlings for 6-10 month growing seasons in seasonal tanks, are now the focus of government and NGO policy. In early studies, bighead carp and tilapias produced the highest levels of output, albeit bighead carp appears to be hampered by its poor marketability. The most ideal species for community production programs are likely to be rohu and the more easily bred common carp, which also did well. Such programs should be carried out at the watershed level to maximize the flexibility of stocking programs in the case of extreme climate change, according to the authors. Uptake is being hampered by a lack of institutional cohesion, a lack of com, and inadequate scale economies associated with seed requirements for small-seasonal tanks.

Alternative low-input solutions based on collecting local seed from perennial systems and seasonal spill events require more research. Where tilapias are stocked, high predation pressure by drought-tolerant air breathers such as the snakehead *Channa striatus*, which is a constraint to culture-based enhancements, is likely to prove advantageous. This prey/predator synergy could reduce overcrowding-related stunting and boost net yields, including a profitable by-catch of snakehead, which is quite popular with consumers. Alternative aquaculture operations in other irrigation subsystems are also taken into account. Due to the high investment requirements involved with purpose-built buildings, as well as the cheap overheads of the highly profitable and neighboring capture fisheries, pond aquaculture in the dry zone is seen to have limited potential. Agro-wells that use ground and seepage water from community tanks may have the capacity to produce advanced fingerlings. A bilateral research program is currently looking into the possibility of advanced fingerling production in flooded paddy fields, however, this is not a practical concept in the context of rainfed tank systems, where storage is typically only adequate for supplementary irrigation. Women's groups have recognized ornamental fisheries, which have a strong export market but require little capital investment, as promising income-generating industries. Once again, marketing constraints are cited as major roadblocks to adoption.

Cultivation of these species may aid in the survival of species that are under a lot of secondary pressure from the tilapia gill-net fishery. With a coast line of 1585 kilometers and a continental shelf area of 26,000 square kilometers, Sri Lanka's primary focus for fisheries development is on marine fisheries(Athukoorala, Bhujel, Krakstad, & Matsuishi, 2021). However, because markets for marine and inland produce are highly spatially separated, inland produce remains vital to the nutrition of some of the poorest sections of society in rural inland areas. Rivers, estuaries, lagoons, brackish water ponds, floodplain lakes, and major, minor, and village tanks are among the inland fisheries resources. From the beginning of recorded history, inland fisheries appear to have played a vital role in Sri Lanka's subsistence and self-sufficiency economy. However, until the recent emergence of a high yielding commercial tilapia fishery following the introduction of the fast-growing foreign cichlid, *Oreochromis mossambicus*, during the 1950s, there was no tradition of aquaculture or organized freshwater fishing. Apart from a recent surge in prawn culture in coastal belts and lagoon edges, which is currently plagued by disease and environmental issues, there is no significant aquaculture activity in the state. However, due to the loss of these productive ecosystems as a result of river valley development operations, man-made reservoirs now account for the vast majority of inland production. Large coastal lagoons around the East and Northwest coasts also sustain highly profitable fisheries, however only in regions where good marine resources are already available.

Figure 7.5: Cyprinus carpio.

Source: https://binary.pensoft.net/fig/421632?size=thumbnail

Despite the enormous number of species transferred over the world, effective introductions have been based on few species. Introduced tilapias have become one of the most important groups of fish for both subsistence and commercial exploitation in Southeast Asia. Capture fisheries and extensive to extremely intensive farming systems are examples of production systems. Tilapias, and to a lesser extent common carp, *Cyprinus carpio*, (Figure 7.5) have contributed to the recent increase in inland fish production in lakes and reservoirs in Madagascar, Mexico, Peru, and Cuba. In 1988, Mexico increased its production from 2,000 to 123,000 metric tonnes per year, with tilapia accounting for 65 percent of the total and common carp accounting for 24 percent(Jiménez-Badillo, 2006). Tilapia and common carp account for the majority of the 40,000 tilapia and common carp produced annually in Madagascar, a tropical island with many parallels to Sri Lanka's circumstances. In 1988, Cuba and Brazil produced 15,000 and 17,000 tilapias, respectively, accounting for 93 percent and 30 percent of total production. The fast expansion of China's aquaculture industry, which is mostly based on indigenous carps, tends to obscure the growing relevance of tilapias. China now produces 47.8% of the world's farmed tilapia, five times more than the Philippines, which is in second place. The exception to this success tale is the Indian subcontinent. Tilapias are often regarded as pest species in this country. They are thought to have spread at the expense of larger-growing, higher-value carp species, which today dominate commercial markets. Because of this viewpoint, it remains technically forbidden in India, although being widely practiced and even cultured in West Bengal, where it is popular among both commercial producers and consumers. It's also essential in the southern tanks. This viewpoint ignores the importance of smaller, less expensive species to impoverished rural customers. Early settlers in Sri Lanka spread along the banks of rivers and streams, and irrigated paddy agriculture became increasingly significant as a result. Because of the limited communications at the time, only a small amount of fresh sea fish could have been brought inland. Fish have been captured in rivers, streams, reservoirs, and irrigation canals instead. According to rock inscriptions, such fishing was common throughout the early Anuradhapura period.

Fishes were the property of village assemblies in the case of communally developed village irrigation works. Smaller tanks were frequently built and owned by individuals. Small ponds built in garden homesteads and waterways that went through or around paddy fields were the most widely owned private water resources. No species with rapid growth and proliferation traits of commercial significance existed in the local fauna, necessitating exotic introductions. Pitipana and Manar Brackish Water Culture Stations were developed in the 1950s. The focus has switched to fisheries enhancements based on the stocking of main and minor tanks with carp and tilapia fingerlings since the mid-1970s. With Chinese and then Indian bilateral technical support, new fingerling production sites were developed to supplement the first plant at Polonnaruwa. O. mossambicus, which is well suited to Sri Lanka's shallow and highly productive perennial reservoirs, filled this artificial lacustrine niche in a way that none of the slower-growing indigenous riverine species could. Sri Lanka presents a unique case in a worldwide spectrum of acceptance for tilapiine enhanced fisheries due to its unique structure of water resources and history of aquatic resource production, but one from which useful regional lessons can be gained. The fishery began as a way for communities living near reservoirs to supplement their agricultural income. Coastal fishermen increased exploitation by migrating to reservoirs during the monsoon season. Capture and culture fisheries are the two types of inland fish production in Sri Lanka, both of which are confined to man-made reservoirs. The culture fishery, which is still in its early stages, is based on annual stocking of tiny seasonal tanks, whereas the capture fishery is focused on bigger perpetual tanks. The disparity in the contribution of these two sectors is evident in the fact that capture fisheries account for about 90% of total commercial inland production in just 74 major reservoirs. Both systems exhibit a high level of natural productivity.

Tilapias are currently split into three general groups, primarily based on their reproductive mechanism. These include substrate spawning, biparental mouth-brooding, and maternal mouth-brooding Tilapia, Sarotherodon, and Oreochromis, respectively. The larger commercial species, such as *P. sarana* and *L. dussemiri*, are rigidly potadromous and must migrate back to their original riverine settings to spawn, making them more vulnerable to pollution and environmental deterioration. Tilapias, clupeids, and many smaller cyprinids have more adaptable breeding and feeding habits than other fish. Tilapias have the ability to regulate their fecundity in response to fishing pressure, which is a significant benefit in the face of ever-increasing exploitation. Because of its superior growth qualities and body conformance, this species has been the preferred species for stocking programs since the late 1970s. Introgressive hybridization with established *O. mossambicus* populations has occurred in most reservoirs since its introduction. *O. mossambicus* has invariably gained admission through inflowing water supplies, even in newly impounded lakes stocked entirely with *O. niloticus*. Furthermore, because the high yields off O. mossambicus are a result of its established resilience to the local climate, complete replacement is unlikely, even in seasonal tanks. Furthermore, while smaller fish command lower prices than larger specimens, they are cheaper and most typically consumed by lower-income groups, raising doubts about the utility of increasing average sizes in an artisanal fishery with an entirely rural consumer base.

While all introduced tilapias can self-recruit, only *C. carpio* is known to have a self-sustaining population among the numerous species of carp introduced into the food fishery. It is presently the only exotic found in headwater streams in the wet zone above 1500 meters above sea level(Pethiyagoda, 1994). Three Chinese carps like *Hypophthalmichthys nobilis*, *H. molitrix*, and *Ctenophyaryngodon idella* and three Indian big carps like *Hypophthalmichthys nobilis*, *H. molitrix*, and *Ctenophyaryngodon idella* are among the six additional introduced species. Only L rohita is known to be capable of spawning at a low level in Sri Lankan waters.

Because of their inability to reproduce naturally in tropical latitudes, all of these species have fallen dramatically since state-sponsored stocking halted in 1990. The common carp, on the other hand, appears to be an outlier. When larger individuals rise up from the tank floor during particular times of the year, they become amenable to gill-net fishing in small amounts. Different cohort sizes were detected in harvests from Rajangane and Usgala Siyambalangamuwa reservoirs in North West Province in the absence of recent stocking programs, implying that limited, low-level recruitment is potentially viable in low land reservoirs. In contrast to the mainly larger 5-20 kg fish seen in NW Province, these were mostly smaller specimens. These were likely the result of more recent stocking attempts by the neighboring Uda Wallawe Government hatchery, as they were too few to be leftovers of older stocking programs. When fish are stocked into waters where natural reproduction occurs, the dynamics of the process become less predictable as a number of density-dependent factors enter the picture, affecting nutrition, growth, mortality, and reproduction. The proportion of supplied fish vs natural reproduction fish will have an impact on these outcomes. Stocking may be unnecessary in circumstances where spontaneous reproduction is sufficient.

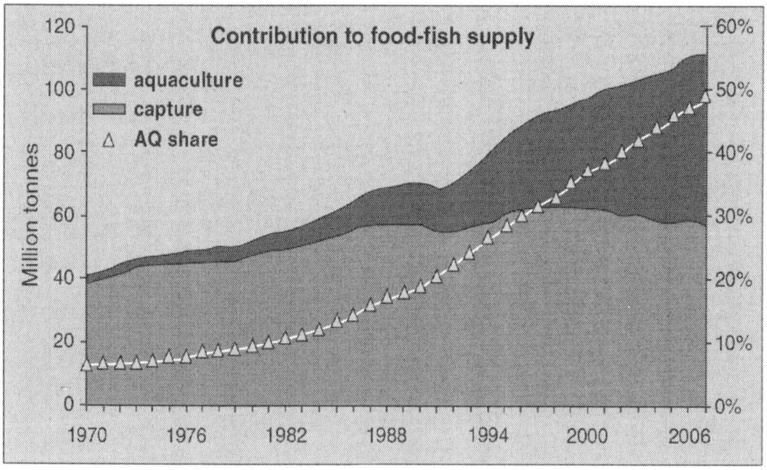

Figure 7.6: Capture fisheries.

Source:https://www.researchgate.net/profile/Upali-Amarasinghe-2/publication/226799090/figure/fig3/AS:644664973590529@1530711515062/Contribution-from-capture-fisheries-and-aquaculture-to-fish-production-and-the-percent_W640.jpg

7.7 CAPTURE FISHERIES

Capture fisheries management (Figure 7.6) in Sri Lanka, as in most other Asian countries, is either non-existent or inadequate. Aside from the stocking of different tilapia kinds over time, the fisheries can be considered unmanaged. Although there are legal restrictions on the sorts of gear that can be used and the amount of fishing that can be done, enforcement and monitoring capacities are lacking. In Sri Lanka, cooperative organizations are ineffective, and disputes with outsiders, who are frequently internally displaced persons, are becoming more widespread. Various scholars have identified trends in output loss as a result of these issues. While fishermen and merchants agreed on long-term trends in species composition, they disagreed on catch volume patterns. Over this time period, all fishermen agreed that the number of fishermen and the intensity of fishing tactics had increased considerably. Rather than the rapid drop recorded, an initial rise in landings could be expected in response to deregulation, followed by a reduction in subsequent years when the impacts of overfishing and stocking were felt. In 1985, it was anticipated that *O. mossambicus* accounted for 70-90 percent of inland landings, but available figures reveal that previous to 1989, a maximum of 10 million mixed species were stocked per year, a small quantity(Amarasinghe, 1987).

7.8 HEALTH

While the exact health of the fishery can only be speculated upon, native species and carp catches have clearly decreased, while prolifically breeding tilapias appear to have managed significantly better in the face of rising exploitation pressure. Several responses claimed that the introduction of common carp ten to fifteen years ago coincided with the rapid decrease of numerous indigenous species, including puntius sarana, ompok bimaculatus, gigantic snakehead, snakehead, and, in particular, the endemic carp species *Labeo dussemiri*. The true interplay of causes responsible for this reduction has to be investigated further, although secondary pressure from the tilapia gillnet fishing, which grew substantially over the same time period, is likely to be one of the most serious culprits. Other environmental issues have also been linked, such as habitat alteration and increased pollution. Agrochemical applications, particularly pesticide applications, have also increased dramatically over this time span, with the most devastating repercussions expected to occur in endemic species' riverine breeding areas. Consumer impressions of various carp species, market conditions, yield potentials, and stakeholder participation in villages where stocking programs have been resumed and existed before the government's ban will all require more research. Seasonal tank fisheries in the arid zone have traditionally been managed to exploit naturally recruited stocks on a subsistence basis in a large-scale but sustainable manner. According to historical data, practically all male members of a village's families used to go fishing on occasion to supplement their nutrition. During the dry season, the ecological setup of small tank irrigation systems aided fish survival. Fish that managed to survive in larger tanks with some water left over could travel upstream to colonize and reproduce in smaller seasonal tanks. Fallen leaves offered nesting materials, hide-outs, and refuge from predators.

Although inefficient, they tended to be size and species selective, allowing juveniles to escape and protecting the fishery from being over-exploited. Only the proprietors of the smallest, sometimes private irrigation tanks have fishing rights. Where tanks were common property resources, only villagers from the village had access to the fisheries within them, which were managed by village assemblies. In contrast, they generally had no access to outside supplies, forcing them to conserve the village's fisheries resources. Similarly, to the irrigation need for water, a family's right to fish was defined by the size of its paddy lands. The practice of collaborative fishing guaranteed that catches were distributed fairly. Fish were primarily caught to supplement household diets, with only a few exceptions. In practical terms, fulfilling these standards entails locating an operating location that satisfies all of the basic requirements at a reasonable cost. Suitable water, land, feed supplies for the fish or shrimp, replacement stock, and support and logistical facilities for processing, shipping, and storage are the most basic requirements.

Rain, surface streams, lakes, ponds, and springs are the most common sources of freshwater. The climate in Sri Lanka's Eastern Province features an extended dry season with inconsistent and erratic rainfall. Many excavated ponds are left dry for over half of the year due to a reliance on rainwater alone to fill them. It is often discouraged to use canal water to fill aquaculture ponds since this would compete directly with irrigation for rice paddy production and other existing agricultural activities. The property adjacent to sea coastlines, lagoons, and estuaries has easy access to marine and brackish water. However, businesses must be cautious in their use of this water to avoid contaminating groundwater with brackish or seawater. Intakes and discharge pipes/canals must also be placed so that they do not obstruct boat landings, storage areas, or other existing operations. The availability of low-cost land is one of the most significant prerequisites for land-based aquaculture. Even fish cage culture necessitates a landing location for boats and other support activities, particularly if value addition is planned near the culture site. Pond farming of fish and shrimp necessitates generally level land near plentiful water sources. Ponds must be built in such a way that they do not leak and contaminate nearby ground water, and that they can be drained and cleaned at regular intervals without harming the environment.

There are further issues of land tenure and competing usage to consider. If the investors have a land title, it is easier to finance an aquaculture venture. Long-term lease agreements are, at the absolute least, required. Land can become too expensive for aquaculture to be economically feasible due to fluctuations in land values. Aquaculture activities are also ideally positioned in places where they do not immediately compete with or interfere with existing agricultural, commercial, or residential activity. For a variety of factors, Sri Lanka has excellent grounds to investigate the prospect of exporting various items, at least to neighboring Middle Eastern countries. To begin with, refrigerated transportation is easily accessible. Dole Asia chose Sri Lanka as the location for its fresh banana export operations because of the low transport periods and large number of vessels that operate between Colombo and Middle Eastern ports. The possibility of increased prices is the reason to investigate export options. Nobody can promise anything about pricing other than the fact that they will fluctuate over time. Fish prices fluctuate a lot from year to year, season to season, and even week to week. Different fish species compete with one another to some extent, therefore increases in one species' supply might affect the market price of others.

7.9 ENTERPRISES

Enterprises interested in practicing large-scale aquaculture in Sri Lanka will also need to secure scientific backing to defend against diseases like those that ravaged Chilean salmon farms and wreaked havoc on Sri Lanka's own prawn and shrimp production. Health organizations are becoming increasingly suspicious of disease-prevention techniques that rely nearly completely on the use of antibiotics. Antibiotics have been found in fish imported into the EU, which has resulted in their rejection. Because of the presence of chloramphenicol, an antibiotic found in imported marine items, the EU banned the import of animal products, including frozen shrimp and prawns, in 2002(Mohamed Abdelrahman, 2018). The World Wildlife Fund (WWF) and the Global Aquaculture Alliance (GAA) are two organizations that have established criteria for tilapia farm certification. Tilapia Best Aquaculture Practices (BAP) have been finalized by the GAA. The BAP standards and recommendations for tilapia farms apply to cage and net pen farming, as well as pond culture, and share many features with the BAP requirements for other farmed species. Regardless of the system in place, farmers must adhere to restrictions governing the use of non-native species. Only 8% of Sri Lanka's population lives in the Eastern Province, which looks to have more land accessible than the Western and Southern Coasts at first glance. First impressions are no guarantee of finding and obtaining the correct land. Land with access to fresh, brackish, or saltwater will be required for successful aquaculture. The quantity of land required to maintain an aquaculturist and his or her family varies depending on the produce.

Large tracts of land may not be readily available or affordable enough to entice investors. Only ornamental fish and plant producers can afford to pay high land costs since they can operate and make enough money on small pieces of land. On the East Coast, land with access to brackish or saltwater looks to be considerably easier and less expensive to come by than in Western Sri Lanka. Land near brackish water is less likely to be utilized for agricultural or industrial purposes. Pond culture of freshwater fish species such as tilapia, milkfish, and carp species have various advantages for newcomers to aquaculture. Although these species have a lower market price than shrimp, they are easier and less expensive to cultivate and are less susceptible to disease. Brackish or saline water, which is more easily available in the Eastern Province than freshwater, can be used to cultivate tilapia and milkfish. Freshwater shrimp should be stocked in ponds with tilapia and carp when freshwater is accessible. Local markets should be targeted for pond-raised fish because these markets are prepared to buy fish without ice or other forms of processing.

CHAPTER 8

TILAPIA ECONOMICS OF THAILAND

CONTENTS

8.1 INTRODUCTION

Thailand is a leading fish producer (Figure 8.1). In 1999, total fisheries production was above 3.6 million tonnes. The relatively large annual fish output can be attributable to a number of factors, one of which is geographic advantages (Harohau et al., 2016)aquaculture is relatively new in the Pacific and such potential is hampered by a lack of aquaculture knowledge and practice within local cultures. A participatory action research approach was used to conduct on-farm trials with farmers in Solomon Islands to develop relevant and improved ways of farming and maximising productivity of the resident exotic tilapia Oreochromis mossambicus. During the 34month period when the research was undertaken improvements were evident, through increased farmer participation and improved knowledge of farmers on pond design and fish husbandry techniques. One of the contributing factors to improved farmer understanding was the production of knowledge products which were co-developed with and based on the farmers' local context. Productivity of a typical 20–38m2 pond ranged from 726 to 1819kgha−1year−1. Because 80% of this production was consumed by households, such ponds, producing easily harvestable small tilapia, have a role in supplementing household diets and contributing to improved nutrition at the subsistence level. While pond system productivity reached levels equivalent to low input tilapia ponds in other regions, the resident tilapia in Solomon Islands has limited opportunity for more than low level commercial enterprises. Established ponds are included in the daily livelihood tasks of both men and women and explicitly gender equitable approaches to partnerships with pond farmers provide opportunities to further increase benefits to households. A participatory action research approach, and the principles embodied therein, is recommended for further development of household aquaculture enterprises, regardless of species, in Solomon Islands.\nStatement of relevance\nWe feel that our paper makes significant and novel advance to the field of aquaculture by:(1. Thailand has a total land area of 540000 km^2 and a 2614 km coastline. The marine fishing grounds that come within Thailand's Exclusive Economic Zones are located in the Gulf of Thailand and the Andaman Sea, and span a total area of approximately 316000 km^2. Inland waterways cover an area of about 3750 km^2 (Yamkasem, Tattiyapong, Gorgoglione, & Surachetpong, 2021)2021.

Figure 8.1: Tilapia in Thailand.

Source:https://www.globalseafood.org/wp-content/uploads/2011/07/BHUJEL-pic1.jpg

Furthermore, coastal aquaculture has the potential to cover approximately one million hectares. The fisheries sector's gross domestic product (GDP) was 123.2 billion Baht in 2000, accounting for around 2.5 percent of national GDP and 27.6 percent of agricultural GDP, respectively (Deines, Wittmann, Deines, & Lodge, 2016). Other associated and supporting sectors, such as ice plants, cold storage, fish processing facilities, shipbuilding, hatcheries, feed industry, transportation, and so on, have all benefited from the development of the fisheries industry. As a result, approximately two million people are employed in this industry, with 40% of them being fishermen and fish growers and the rest working in connected or supporting businesses. The fish produced is either consumed locally or exported to gain foreign income. It is an essential protein source, as evidenced by per capita fish consumption of 25-32 kg during the last decade. The value of fish and fisheries goods exported has increased dramatically. In 2001, the country had a trade surplus of 146.8 billion Baht. Thailand became the world's largest fish exporter in 1993, and the country's ability to meet expanding fish demand depends on the balance between supply and demand. The significance of international fish commerce in terms of generating money, employment, and export earnings, all of which contribute to the nation's prosperity need to be understood. Per capita fish consumption and expenditure can be examined, with geographies and consumer types defined. Thailand's fisheries productivity has increased dramatically over the last three decades. In 1977, total output surpassed two million tons for the first time, but it was followed by a setback, and it has not surpassed two million tons since 1982. In 1999, approximately 3.6 million tonnes were recorded, with marine capture fisheries accounting for 75.2 percent of the total (Yi, Lin, & Diana, 1996).

8.2 CONSUMPTION

The amount of fish consumed by urban residents in inland regions was only 19.92 kg, which was significantly less than the amount consumed by rural people. Rural producers consume 34.92 kg of fish per year, whereas rural non-producers consume 28.68 kg. These numbers show how important fish is as a protein source for rural consumers to meet their nutritional needs. In urban regions, consumers may have access to a number of additional animal protein foods. The urban consumption of fish in coastal provinces was 42.59 kg, which was lower than the 53.56 kg consumed by rural producers, but higher than the 40.74 kg consumed by rural non-producers. For all consumer groups, fish consumption is higher in coastal provinces than in inland provinces (Madan M. Dey et al., 2000). The fishing population, in particular, benefits from easy access to fish supplies in coastal areas. In inland provinces, the species composition of the fish basket differs for rural and urban consumers. Tilapia is the most popular fish across all consumer groups. Snakehead comes in second place for both rural non-producers and urban customers, whereas silver barb comes in second place for rural producers just after tilapia. The species composition in coastal provinces is consistent for all types of customers. The Nile tilapia is a freshwater fish with several benefits over other culture fish, including rapid growth, the ability to use a variety of diets, tolerance for high stocking densities, and a wide range of environmental conditions. In addition, tilapia consumption is on the rise all over the world. As a result, tilapia is vital to the global economy. Furthermore, clay ponds are used to cultivate 79.9% of Thailand's tilapia culture (Thunjai, Boyd, & Boonyaratpalin, 2004). The remainder is raised in floating baskets, rice fields, and orchard water canals. Commercial tilapia farming in Thailand is divided into four stages: fry, fingerling, juvenile, and adult fish farming. Most farmers who raise tilapia in floating baskets employ juvenile tilapia rather than breeding their own fry to raise adult fish.

8.3 COMMERCIAL FISH FARMING

Medium and large-scale commercial fish farmers, on the other hand, must employ the larger ponds to produce vast quantities of fish. Despite this, there is a scarcity of knowledge on the impact of pond size on tilapia juvenile production and economic profitability. The vast disparity between communities with adequate access to fish and those without is hidden by this national average. The broad range of fish consumption reflects the large range of income. In isolated areas in Northern Thailand, seafood consumption is very low, averaging 3–5 kg per capita. These may not, however, include fish obtained and consumed outside of the hamlet. Thailand, with a land size of about 514,000 square kilometers, is located on Southeast Asia's Indochina Peninsula (Tain & Diana, 2007). Myanmar borders it on the west and northwest, the Lao People's Democratic Republic on the north and northeast, Cambodia on the east, and Malaysia on the south. The climate of the country is monsoon type, with distinct wet and dry seasons. Except in the south, where there is no obvious cool season, the rainy season runs from May to October, the cool dry season from November to February, and the hot dry season from March to May. The yearly precipitation ranges from 760 to 4,200 millimeters. Droughts and floods are prevalent in Thailand, especially in the northeast. National development plans, particularly the construction of multifunctional dams, have substantially diminished the value of floodplain fisheries. Although these dams provided reservoirs for water storage for agriculture and/or energy generation, they also reduced flooding in floodplains and places that functioned as natural spawning and nursing sites for most fish species. Fish populations and catches have plummeted as a result of this, combined with pollution and environmental degradation. Thailand's freshwater fish fauna is diverse due to the country's extensive network of rivers and canals, particularly in the central plains, as well as numerous swamps, reservoirs, and water storage tanks. Freshwater fisheries, on the other hand, have declined, providing a major stimulus for the country's relatively recent development of aquaculture.

8.4 GEOGRAPHICAL CONSIDERATIONS

Geographical considerations and Thailand's fish culture tradition influence aquaculture productivity in the country. Thailand has over 120,000 km of riverine systems, 300,000 ha of natural lakes, and 255,000 ha of reservoirs. Thailand has a wide range of freshwater aquaculture methods and species, including several that are important to small-scale farmers and the country's food security. More than 20 fish species are farmed, according to DOF figures, with total freshwater fish production of 271,000 tons in 2000 (Dan & Little, 2000). Despite the fact that data is not organized by scale or intensity of fish farming, herbivorous and omnivorous fish species with the greatest relevance for small-scale aquaculture such as carps, gouramis, and tilapias make up roughly 60% of the total. Aquaculture may have begun in Thailand as early as 1691, however, it was for beautiful goldfish rather than food. Because of the abundance of wild fish, aquaculture for food fish looks to be a comparatively new development. The native riverine catfish has been farmed on a modest scale in cages and ponds in Central Thailand. Using fish fry supplied by boat from the People's Republic of China, it is widely assumed that Chinese immigrants introduced organized aquaculture to Thailand in the early years of the twentieth century. Chinese carp and grass carp were farmed on a modest scale, especially in Bangkok, where the huge immigrant Chinese population provided a ready market.

DOF began researching the native snakeskin gourami and its life cycle in 1932, despite the fact that its entire life cycle takes place in flooded rice fields, it was unable to reproduce. It was possible to persuade people to farm it because wild fish were still plentiful. The 1950s were a pivotal decade for women's liberation. The FAO provided technical help. The UN Department of Foreign Affairs imported and disseminated Mozambique tilapia (*Oreochromis mossambicus*). It became a popular farmed fish in ponds very quickly. However, the popularity of this species has faded. Due to concerns about the quality and flavor of its flesh Later, tilapia farming grew in popularity. After His Majesty's introduction of Nile tilapia (*Oreochromis niloticus*) in 1965, it was established. His Imperial Highness Emperor Akihito of Japan presented King Bhumipol with several specimens as a gift. The farming of snakeskin gourami is an exception. Farmers converted barren rice fields into moderately saline soils southeast of Bangkok. Riverine

catfish, silver barb, and common carp. The availability of these fish species' seeds resulted in a significant boost in aquaculture production (Na-Nakorn, 1992).

8.5 SMALL-SCALE AQUACULTURE

For small-scale aquaculture, tilapia and herbivorous and omnivorous carps are generally farmed in polyculture in ponds. Small-scale farmers are constrained by limited on-farm sources of fertilizers and feeds, as well as access to, and affordability of, marketed feed, according to the findings of a survey of fish farms conducted more than two decades ago in Pathumthani Province in Central Thailand to identify aquaculture techniques for small-scale farmers. Large commercial fish farms are either integrated with animals or use waste food from factory canteens and restaurants, as well as diverse agro-industrial factory by-products. Most integrated fish farmers are essentially livestock owners who have built ponds in the floodplain to raise the level of their animal quarters during the rainy season to protect animals from drowning. Livestock quarters are situated above or adjacent to the pond, allowing manure to fertilize the fishponds and spilt feed to offer nutritional inputs to the fish. Integrated farms continue to supply the majority of low-value fish to urban consumers. Small-scale farmers have little interest in these types of aquacultures, which are dominated by entrepreneurs in peri-urban regions, particularly in the provinces of Central Thailand. Through research collaborations between the Asian Institute of Technology and the Department of Agriculture, many of the technical challenges faced by new entrant small-scale farmers have been addressed. The majority of the research was conducted in resource-scarce North-eastern Thailand. Although much of Thai aquaculture is concentrated in intensive farming systems, several small-scale aquaculture systems have emerged in former rice fields in recent decades, including snakeskin gourami farming, Nile tilapia seed production, giant freshwater prawn farming, and inland culture of shrimp. Most farmers who raise these species have improved their fortunes; many were impoverished rice farmers before turning to aquaculture (Ben Belton, Little, & Grady, 2009).

8.6 MARKETING

Freshwater fish marketing in the United States is complicated, including a variety of routes, markets, and parties. Freshwater fish distribution and marketing are generally efficient, and market access for fish farmers, especially small-scale producers, is not a problem. With the exception of a state-owned firm, the Fish Marketing Organization, which manages a Bangkok assembly market, fish marketing is primarily in the hands of the private sector. Auctions and negotiations are used in assembly markets to trade fish. Freshwater fish can be easily delivered from production sites to markets across the country since postharvest support facilities are adequate. Fish transportation is aided by well-maintained road networks that connect all districts and provinces. Freshwater fish prices are competitive due to a large number of fish farmers and suppliers, as well as a large number of purchasers along the fish marketing chain (Wati, Chang, & Mustadjab, 2013).

Central Thailand has lower rates than the northeast and south. The snakehead is the most valuable. For consumption, rural markets prefer small tilapia, but urban markets demand larger fish. Because the time lag between harvest, marketing, and final sale to consumers is brief, price changes at the farm gate and wholesalers are harmonized. Fish prices are generally higher from January to September than in other months. Freshwater fish are available for purchase both living and dead. Sand goby, snakehead, and walking catfish are more likely to be sold alive than less valuable herbivorous and omnivorous fish, which are frequently sold on ice. Live fish commands a price premium of up to 50% over iced or cooled fish. In rural places, little fish are marketed to low-income clients. Small-scale fish farms produce the majority of these fish. Large fish are generally provided by commercial cage and fishpond operators, which are chosen by relatively wealthy consumers and restaurants. Although some farmers transport and retail fish themselves, most fish are sold to fish agents at the farm gate when they are harvested. Wholesale markets are primarily found in major cities, and many of them also have retail stores. Wholesalers carry fish to wholesale markets early in the morning, either directly from farms or from assembly markets, for distribution to retailers. Retail markets can be found across metropolitan and peri-urban areas, with many stalls on a concrete floor with power and running water inside a conventional open hall. Freshwater fish are sold live or dead, whole or in pieces, and fresh or processed, such as dried, salted, minced, fermented, or turned into fish balls, in retail markets.

8.7 FRESHWATER FISH

Freshwater fish is also sold in supermarkets with a 40–50 percent markup above normal retail pricing. Wholesalers also ship chilled and frozen desirable species such as marble goby and low-value riverine catfish to neighboring countries. Carnivorous, herbivorous, and omnivorous species including mrigal, rohu, silver barb, and tilapia are also exported to the Middle East by processors and cold storage operators, as well as riverine catfish to Asia and Europe. The marketing margin, or the gap between what the consumer pays and what the producer receives, varies by species. For silver barb and tilapia, the farmers' portion of the retail price can be as high as 50% and 60%, respectively. Tilapia's 40 percent marketing margin is made up of 16 percent direct marketing costs and the remaining 24 percent gross margins divided among fish collectors, wholesalers, and retailers. Retailers cover the costs of cleaning, gutting, and cutting fish before selling it. Retailers also run the risk of not selling their seafood fresh, which can cut retail prices dramatically.

While there are fisheries groups or cooperatives in places where there are a lot of aquacultures, the duties of these farmers' organizations are mostly focused on marketing, which is a common interest among farmers. Because of many inadequacies affecting farmers' groups and aquaculture extension services, DOF has been unable to marshal the assistance of these organizations to deliver its extension initiatives. Freshwater aquaculture cooperatives that are privately operated have always experienced financial and human resource constraints. Community participation in aquaculture development has been seen through village committees, district councils, and subdistrict TAOs. Fish farmers and villagers take part in planning and making decisions about their community's resource usage and conservation in this environment. The government is working to make the extension system more responsive to farmers' actual needs, especially by giving information that is more relevant to their circumstances.

Village schools in distant places, primarily elementary and, to a lesser extent, secondary schools, are the target areas. This program's major goal is to improve the nutritional status of school children in these communities by providing fish for consumption through self-help fish farming projects. Construction of fishponds, aquaculture training, and the giving of fish seed and technical guidance to schools are all part of the program, which began in 1992. In order to improve fish production at a reasonable cost, the Lunch Program has also piloted an integrated fish-poultry farming initiative. Despite the positive results, the program is hampered by a lack of water, inadequate feed and other inputs, and a lack of expertise in fish farming. There is a need for improvement and expansion in order to take advantage of village schools as institutions that serve as a grassroots focal point in rural places. In fishpond management and other aquaculture-related concerns, there are chances for adaptive and hands-on learning. Students and communities can participate in an experiential learning method that actively displays the potential benefits of improved fishpond management to livelihoods and human nutrition through a series of simple activities utilizing village fishponds or small waterbodies. Thailand's National Fisheries Policy on Aquaculture aims to increase fish production to meet domestic demand and increase income for fish farmers. Current strategies include conducting research on fish species with the high economic potential to improve their desirable characteristics and develop good practices for hatcheries and aquaculture farms; providing technical services and certifying registered hatcheries and farms; and providing technical services and certifying registered hatcheries and farms.

The National Fisheries Policy is predicated on the idea that future rural aquaculture development will be small-scale and subsistence-oriented, primarily for domestic use and local household food security, particularly for the rural poor. This restricts the possibility for system improvement. Researchers' main task is to come up with novel and viable low-cost, low-input technology solutions for such situations. In 2001, the government delegated control of fishing resources in all community waterways to TAOs. TAOs have previously aided aquaculture growth in their jurisdictions by asking government funding for fishpond construction and receiving technical assistance from the Department of the Interior. These fishpond development programs have varying degrees of success. Inadequate fishpond management, insufficient extension services, deficient co-management methods and practices for common and shared assets, and limited access to aquaculture information at the village level were the major flaws. The TAOs still lack experience in natural resource management and must establish their credibility and the trust of the communities. Previously, communities lacked the knowledge and expertise needed to make acceptable and enforceable resource management decisions. TAOs can, and increasingly must, play a role in resolving disputes over community natural resources and enabling their management. Through a participatory learning process and iterative improvements, there are chances for capacity building and developing close collaborations between community stakeholders and government services, including fisheries officers and TAO officials.

The Fisheries Act of 1947 now prohibits the development of private ponds in the public domain. Fish farmers, on the other hand, have the right to build fishponds on their own land. Cage culture can also be carried out in public waters by fish farmers. For example, non-obstruction of rivers or transit, non-disturbance to the public, a suitable location, and approval by district and provincial authorities are all necessary for getting government clearance. Fish cage farming licenses are typically awarded for a period of five years. Sub District administrations, the Royal Irrigation Department, the Royal Forest Department, and Thailand's Electricity Generation Department are all currently active in approving cage culture in their jurisdictions. Freshwater aquaculture activities on private property are not required to register or receive approval under the Fisheries Act. Nonetheless, before starting, all aquaculture operators must register with the appropriate authority and obtain a license (Tandavanitj, Henocque, Silpachai, Pimoljinda, & Nabhitabhata, n.d.).

8.8 ZONING

Aquaculture zoning can be used to design and conduct aquaculture activities in order to reduce negative environmental impacts. The fast growth of sea shrimp farms into freshwater areas of numerous provinces in Central Thailand, for example, has resulted in conflicts over land and water resources due to the lack of zoning. Shrimp aquaculture was blamed for salinity intrusion, which harmed freshwater habitats, rice fields, and orchards (Chalermwat, Szuster, & Flaherty, 2003)and during this time, traditional culture techniques have gradually given way to more sophisticated and capital intensive methods. Farmed shellfish production increased from 73,976 to 138,202 metric tonnes between 1988 and 2000. Major species currently under cultivation include the green mussel Perna viridis, the blood cockle Anadara granosa, and three species of oyster (Saccostrea cucullata, Crassostrea belcheri, and Crassostrea iredalei. As a result of this issue, Article 9 of the Environmental Act of 1996 was enacted in December 1997, prohibiting low-salinity shrimp from being grown in freshwater areas across the country. Integrated agriculture-aquaculture has been practiced in Thailand for nearly a century, first in Bangkok and now all over the country. Fish and poultry culture, fish and pig culture, and mixed culture are the most popular systems. The Department of Agriculture has run a number of schemes to boost fish productivity through integrated farming. Because livestock manure is used as organic fertilizer for the fishponds, which also serve as waste stabilization ponds, integrated livestock/fish farming systems protect the environment. Traditional semi-intensive aquaculture employing on-farm and locally available agricultural leftovers has been replaced by a relatively high-cost intensive culture using designed pellet feed over the last decade, resulting in a technology divide.

Intentional and unintentional introductions and transfers of alien aquatic organisms have occurred. Alien species were introduced primarily for aquaculture and the aquarium trade, and many were brought in illegally and without proper quarantine. Freshwater aquaculture is always at risk of negative consequences from the introduction of alien species and farmed fishes, including diseases and parasites. It is necessary to establish enforceable and effective protections that take into account practical biosafety guidelines. However, aquaculture health management guidelines for transboundary movements of live aquatic animals must be implemented such as health certification, quarantine, and diagnostic procedures. Fish farming has grown significantly in recent decades, in part as a result of declining capture fisheries and increased demand for fish. Small-scale farmers have benefited from the development of aquaculture, yet there is no way to quantify the socioeconomic advantages to these farmers based on current statistics. Fish is an important part of the Thai diet, and it helps to provide national food security and human nutrition. Traditional sources of animal protein, fatty acids, and minerals include fish.

8.9 PRODUCTS

Fish and fish products travel readily throughout the country because of good road networks, transit, supporting infrastructure, and telecommunications. This allows Central Thailand to deliver fish to places in need, such as North-eastern Thailand, where retail fish costs are often higher than elsewhere in Thailand. The majority of small-scale farmers in Thailand live in north-eastern Thailand, and these farmers are under increasing pressure to increase agricultural productivity and lower production costs in order to remain competitive in a free market system. Producers, customers, and labor suppliers make up the rural poor. The increased opportunity cost of labor in small-scale aquaculture as a result of rapid economic development and employment alternatives in Thailand and abroad has put further pressure on farm productivity, which must rise if fish farming is to remain a viable livelihood option. The suitability of technology for small-scale aquaculture is constrained by the current economic climate. Low-cost and accessible technology does not always yield high labor returns, and intensive farming might create a demand for financial and other resources that the poor lack. Labor migration to urban regions, particularly Bangkok, has resulted in farm labor shortages in many parts of North-eastern Thailand. Farm households are unable to embrace labor-intensive farming techniques because of these conditions. Over the last few decades, DOF has played a significant role in the development of aquaculture in the country. DOF attempted to encourage native snakeskin gourami farming in the 1930s but failed because of a lack of demand for farmed fish at the time, as well as an abundance of and high demand for wild fish. In the 1950s, the Mozambique tilapia was pushed, but its cultivation failed due to negative traits that limited on-farm productivity and the species failing to match customer tastes and preferences. However, Nile tilapia promotion has been a success since the 1960s, and it is currently one of Thailand's most popular farmed fish species. These development milestones have highlighted the significance of demand, market conditions, and the product's ability to meet demand. The introduction and promotion of fish species for farming necessitate a thorough analysis of demand as well as the ability to maintain a sustainable supply. Nile tilapia fits the needs of Thai consumers as well as farmers who want to produce fish at a low cost and earn a good profit.

The VFPDP is the major tool utilized by DOF to eliminate poverty through aquaculture, particularly in North-eastern Thailand. The program, which aims to boost fish output through village-level community-based management of natural, modified, or constructed water bodies, has had mixed results (Chumnongsittathum, 2008). However, the recent devolution of authority for natural resource management, including fisheries, to sub district governments has opened up new possibilities for enhancing community engagement in decision-making about natural resource usage. With the help of the Swedish International Development Cooperation Agency and the Asian Institute of Technology, the VFPDP program has recently entered a new phase. DOF is actively assisting the rural poor with aquaculture extension services based on a distance extension method, using technologies suitable for household-level and pond-based aquaculture. Because most local communities and individual farming households have limited resources, reaching poor target groups requires less technical but demand-driven approaches. Nonetheless, in the continuously changing rural economy, obstacles in generating sustainable aquaculture technology solutions continue to emerge. Several factors should be considered when responding to challenges to make aquaculture beneficial to small-scale farmers. Capacity enhancement of local government agencies and local service providers is required to meet the difficulties. Without relying on strictly predetermined technological packages, adaptable approaches are required. Analyzing the characteristics of households or small-scale farmers, as well as evaluating the specific peculiarities of their operational environment, are crucial steps in determining how to make aquaculture work for small-scale farmers. According to the DoF database, at least 18 aquatic families were being farmed. The Penaeidae family, specifically whiteleg shrimp and giant tiger prawn, contributed the most to national aquaculture production, followed by the Mytilidae family including green mussel, *Perna viridis*, and horse mussel, *Musculus senhousia* at 15 percent and the Cichlidae family. From 1995 to 2001, the most prevalent species was gigantic tiger prawns, followed by green mussels from 2002 to 2004 (Menasveta, Piyatiratitivorakul, Rungsupa, Moree, & Fast, 1993). Whiteleg shrimps have become the most abundant species since then. Shrimp yields were always the largest contributor to coastal aquaculture production, with mollusks and fish accounting for 36 percent and 2 percent, respectively. The whiteleg shrimp (*Penaeus vannamei*) (Figure 8.2) and gigantic tiger prawn were the most commonly cultivated species (*Penaeus monodon*) (Figure 8.3).

Between 1995 and 2015, the number of farms producing farmed mollusks climbed from 3500 to 6000. Green mussels (*Perna viridis*), blood cockles (*Anadara* spp.), and oysters *(Saccostrea cucullata*, *Crassostrea belcheri*, and *Crassostrea iredalei*) were the most commonly grown species. In 2015, around 16000 acres of land in the Gulf of Thailand and Andaman Sea coasts were used to support shellfish culture. Mollusks are typically farmed around coastlines, where wild or hatchery-reared seeds are grown on the seabed or suspended in nets, ropes, wood, or other structures. Deep-water pound nets and shallow-water pound nets in Thailand's coastal seas accounted for almost 20% of total cultured shellfish harvest by weight in 2015, valued at about US$5.8 million. Eutrophication, which is produced by an overabundance of nutrients such as phosphorus and nitrogen, is well recognized as a serious environmental hazard.

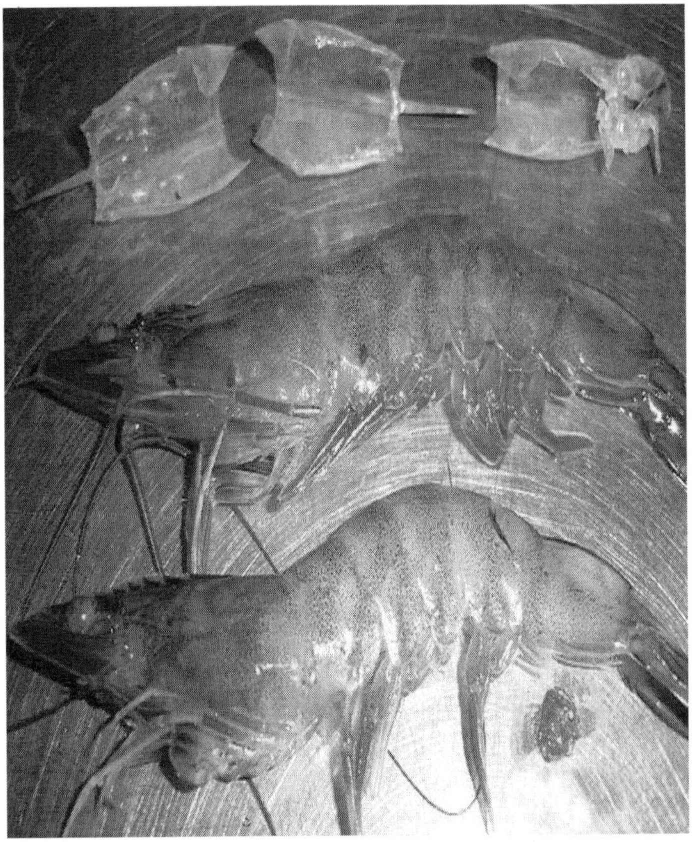

Figure 8.2: Penaeus vannamei.

Source: https://upload.wikimedia.org/wikipedia/commons/thumb/1/1a/Litopenaeus_vannamei_specimen.jpg/800px-Litopenaeus_vannamei_specimen.jpg

It has a deleterious impact on water quality and eventually causes ecological harm. The release of untreated wastewater and sewage sludge from fish and shrimp farms is a major cause of eutrophication, mostly due to the intensification of aquaculture output. Because of the breakdown of waste feed and fish feces, the water quality of overstocked and/or overfed fish farms is frequently poor, and its discharge can have detrimental effects on nearby water sources. Effluents from such farms dump a lot of nutrients into coastal and estuary waterways, which often stimulates the rapid growth of primary producers like algae and plankton in aquatic ecosystems. The blue-green algae flourished in the Gulf of Thailand due to the dumping of untreated sewage.

The issue of wastewater discharge from aquaculture farms has received a lot of attention. Researchers have studied approaches for ecologically friendly aquaculture to reduce nitrogen and phosphorus inputs from point-source effluents to water bodies in recent decades. Biofloc technology is gaining traction as a cost-effective alternative to traditional water management systems. It combines nutrient removal from water with the generation of microbial biomass, which can be used as feed supplements by culture species in situ.

Figure 8.3: Penaeus monodon.

Source: https://upload.wikimedia.org/wikipedia/commons/thumb/2/2e/CSIRO_ScienceImage_2992_The_Giant_Tiger_Prawn.jpg/800px-CSIRO_ScienceImage_2992_The_Giant_Tiger_Prawn.jpg

Furthermore, the concept and practice of integrated multi-trophic aquaculture is one method of decreasing water pollution problems associated with aquaculture. Multi-trophic aquaculture is founded on the idea that one species' waste, including uneaten feed, excrement, and metabolic emissions, is beneficial to the growth of other species, resulting in a natural self-cleaning process. Many nations, including the Philippines, Malaysia, Vietnam, China, and Thailand, have adopted this method of growing fish species in combination with seaweed to boost economic benefits while reducing negative environmental impacts from aquaculture.

Aquaculture production gives a chance to enhance seafood output in the face of population growth, increasing demand for animal protein, and the limitations of expanding wild fishing harvests. Thai aquaculture production has grown quickly in recent decades, accounting for the majority of the growth in fish supply yield. The promotion of aquaculture production has become one of Thailand's core goals, with the goal of ensuring food security while also growing the country's economy. However, concerns about environmental degradation have brought this industry under attention. The Agricultural Standards Act B.E. 2551 (2008), the Thai Agricultural Standard on Good Aquaculture Practices for Marine Shrimp (TAS 7401-2014), and the shrimp Code of Conduct are examples of government efforts to improve and promote a sustainable farming industry through the reformation of Thai aquaculture (Booncharoen & Anal, 2021). Government agencies are assisting in the development of new aquaculture technologies and tools, which are then disseminated to farmers in order to encourage sustainable aquaculture practices. Shrimp farming is gaining popularity in Thailand because of favorable geographical conditions and recent technological advancements that have increased productivity. As a result, the total land area used for shrimp farming has increased dramatically in recent decades. It was stated that coastal locations suitable for marine aquaculture in many nations might meet projected seafood demands, particularly mollusk production. Though aquaculture production in the ocean has the potential to develop in the future, environmentally sensitive or high biodiversity regions, such as coral reefs, should be protected from farming companies. The expansion of future agriculture must be factored into the development of ports and harbors for accessing seafood markets and farming infrastructure.

8.10 PRODUCTION TECHNIQUES

Though aquaculture production techniques can help people get food and nourishment while also helping to grow the economy, an unsustainable expansion of the business poses a serious threat to ocean resources, coastal resources, and the global environment. The large volume of wild fish, particularly garbage fish, required to feed the farmed fish and shellfish industries is becoming a major issue. Alternative protein sources like algal meal, wheat gluten, maize gluten, and insects have been examined in several studies to replace and reduce the use of fishmeal and fish oil in aquafeed production. Furthermore, despite the fact that water quality and quantity are critical for aquaculture output, it appears that adequate water resource management for long-term aquaculture in Thailand has remained a serious concern. Low- and high-tech farming approaches designed for eco-friendly aquaculture, such as the integration of cultures from different trophic levels, rice-fish farming integration, and production systems integration with livestock and agriculture, can be effective solutions to this problem. Innovative methods, such as microbial management of farming systems, are also significant because they can provide a mix of environmental remediation, economic rewards, and social acceptability. Extensive aquaculture may be the most sustainable alternative in some instances, where reduced food production is balanced by other ecosystem services provided by aquaculture ponds. Surprisingly, Thailand's new intense shrimp farming approach is an excellent example of sustainable aquaculture practices. By recycling effluent from shrimp ponds into ponds stocked with tilapia or Caulerpa seaweed, this approach achieves a zero-water-exchange system. Future legislation and research should therefore concentrate on establishing simple-to-implement sustainable aquaculture practices and communicating such information and technology to farmers. Finally, natural disasters such as tsunamis, flooding, and animal disease outbreaks can wreak havoc on aquaculture. Disease outbreaks, such as the early mortality syndrome/acute hepatopancreatic necrosis disease, have impacted Thailand's shrimp aquaculture productivity in recent years.

CHAPTER 9

EFFECT OF CLIMATE CHANGE ON TILAPIA AQUACULTURE

CONTENTS

9.1 INTRODUCTION

The climate has a significant impact on aquaculture production. It is important to learn how farmers raise tilapia in earthen ponds to adapt to climate change (Figure 9.1). Extreme temperatures, severe rainfall, prolonged cloud cover, flood, and drought have all been cited as significant climate-related hazards. To identify critical elements impacting farm revenues, losses, and the choice to increase or decrease ponds, site visits and data collection utilizing questionnaires can be done. Knowledge of fish markets and weather, as well as the effects of weather and climate on fish development, disease outbreaks, and pond water quality, can be given special consideration. Climate and weather-related effects on fish farms at various elevations can be recorded.

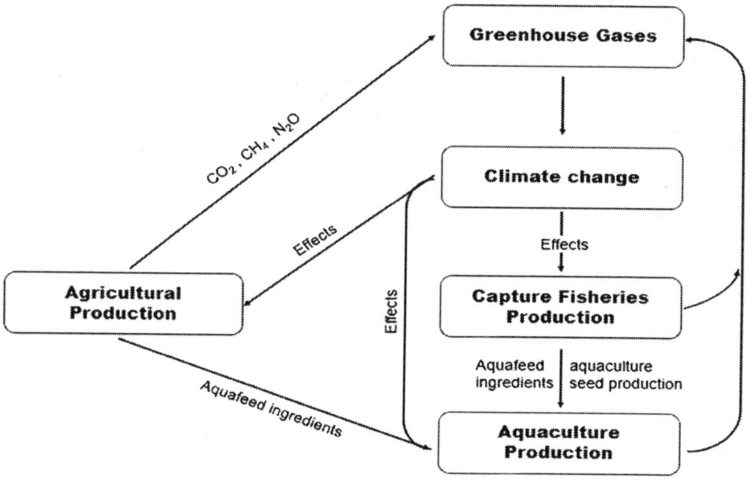

Figure 9.1: Climate change and aquaculture.

Source:https://www.frontiersin.org/files/Articles/609097/fsufs-05-609097-HTML/image_m/fsufs-05-609097-g001.jpg

Farmers in areas where the elevation above sea level is less than 100 meters can be more affected by floods and excessive heat, which resulted in fish mortality and stress, reducing feeding and growth rates. Drought and cold weather, on the other hand, impacted fish farmers in areas where the height above sea level is greater than 500 feet. Disease outbreaks and reduced feeding rates were also a result of these conditions. Farmers responded by limiting feed supply and considering non-fish pond or non-farm supplementary activities as a technique for adaptability. Farmers most typically highlighted high feed prices as a critical issue among nonclimate related variables. The disparities between sites at higher and lower elevations revealed the kind of hazards that farmers may encounter as a result of climate change, which could aid in the development of adaptation plans for individual farmers and the sector as a whole(Rahman et al., 2021).

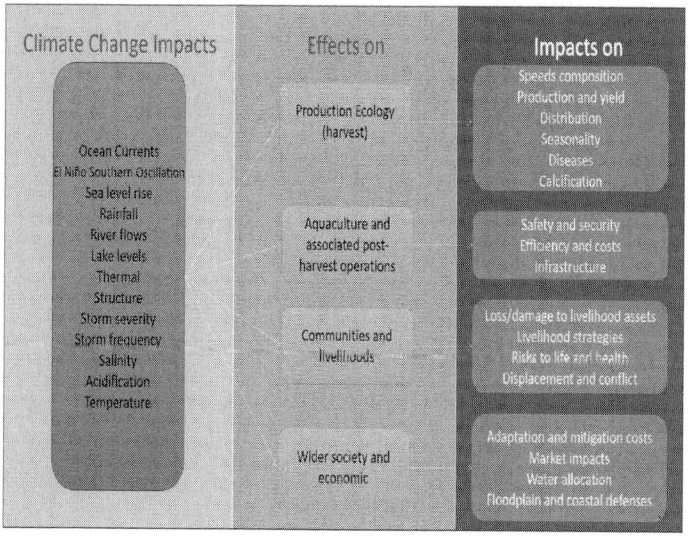

Figure 9.2: Climate change hazards.

Source: https://www.researchgate.net/profile/Jose-Aguilar-Manjarrez/publication/331311343/figure/fig10/AS:730169916526594@1551097482360/Climate-change-impact-pathways-in-aquaculture_W640.jpg

9.2 CLIMATE CHANGE

Recent climate change has been more rapid than in previous periods. Reduced quantities of dissolved oxygen and increased toxicity caused by pollution are two direct effects of rising temperatures on water bodies. Production, on the other hand, may be hampered by environmental agents that damage farmers. In coastal areas like Bangladesh, poor fish farmers are subject to storm surges, excessive rain, drought, saline intrusion, and other environmental risks. Climate change hazards (Figure 9.2) are projected to exacerbate poverty among fish farmers in nations that rely on fisheries and must fight to adapt to climate change. Along Bangladesh's coastal zone, changing climate will result in river bank damage, salinity intrusion, inundation, infrastructure impairment, crop production disaster, fisheries resource destruction, biodiversity destruction, and so on. Furthermore, a one-meter rise in sea level will wreak havoc on the country's vast seashore and flood plain areas, jeopardizing development ambitions and putting environmental refugees at risk. In coastal districts, a large number of poor people work in aquaculture and fisheries. Climate change poses a threat to the residents of these locations. Storm surges caused by flooding, heavy rain, saline intrusion, droughts, eutrophication in stagnant water bodies, and other climate-related dangers are among the most serious.

9.3 HAZARDS

Climate hazards have the potential to reduce the marketing value chain. When fish producers suffer issues as a result of climate change at various stages of production, such as selling, their revenues decline. It was discovered that most of the respondents had experienced problems as a result of storm surges. In Bangladesh's Sundarbans, which are located on the country's south-west coast, events such as cyclones and littoral flooding are considered severe dangers, while storm surges and land erosion are considered 'high risk. Because saline intrusion occurs as a result of flooding, surges, and tropical cyclones, it was not perceived as a threat by poor farmers. Storms, tropical cyclones, and other weather systems, fueled by additional moisture, create more severe precipitation, which is widely observed, even in areas where overall precipitation is declining. Tilapia is a hard species that can be found throughout the year once seeds have been planted(Canevari-Luzardo, Berkhout, & Pelling, 2020). Climate change has an impact on the transportation of fish seed to ponds and matured fish to markets, both of which have increased producer costs. Excessive rainfall floods the ponds, damaging the dykes and causing overflows in ponds without dykes. Because Tilapia is a resilient crop, climate change may not have an impact on its production, but growers are experiencing difficulties as a result of climatic vulnerabilities.

Climate change causes weather variations such as temperature swings, erratic rainfall, and storms. These changes have an impact on aquaculture water quality, resulting in slower growth, increased fish mortality, disease outbreaks, and decreased productivity. Thailand's temperature has been rising for the past 58 years(Chitmanat, Lebel, Whangchai, Promya, & Lebel, 2016). Thailand's average annual air temperature increased by 0.95 degrees Celsius, greater than the average yearly air temperature of other countries, while rainfall is decreasing. The amount and distribution of rain are affected by tropical cyclones, depressions, and typhoons. The number of storms has decreased during the last 56 years, but storm severity has increased. Weather variability, especially in terms of temperature and rainfall patterns, has a greater impact on water quality. As a result, fish growers should be more concerned about climate change. The greatest significant weather variability that affects fish output loss is hot weather. It resulted in fish death and disease. The average temperature was high between March and May, according to statistics from provincial meteorological agencies. The greater temperature in the river had resulted in more water evaporation as a result of the higher temperature. As a result, there was insufficient water to raise fish. Rainfall data shows that rainfall was low from October to March, resulting in a low river flow. The majority of fish farmers considered that the risk of loss was greatest between March and May, when the temperature was high and the amount of water in the river was low and moved too slowly, as well as low flows and dissolved oxygen in the water.

Figure 9.3: Drought.

Source:https://d178ivhysawugh.cloudfront.net/media/l/920x561/1643384496/a-fish-out-of-water.jpg

9.4 DROUGHT

Drought (Figure 9.3) was also linked to the high temperatures. Low water circulation resulted in low dissolved oxygen levels, which led to disease outbreaks and fish death. Dissolved oxygen is required for aquatic respiration and serves as a barometer of water quality. For aquaculture, dissolved oxygen levels greater than 5 mg/L are required. The survival rate of aquatic species is affected by dissolved oxygen levels below 1 mg/L. Fish farmers shifted the cages to a deeper portion of the river, utilized aerators and water pumps, and delayed or postponed the stocking date to lessen the chance of losses. These methods were comparable to those used by fish farmers raising tilapia in floating cages on the Ping and Nan rivers in northern Thailand. By circulating air through the water and creating little bubbles, an aerator and water pump can assist increase dissolved oxygen in water. Fish farmers should also postpone stocking until July when the volume of water in the Songkram river is sufficient for fish cage farming due to the rainy season. Flooding also resulted in fish deaths, disease outbreaks, cage damage, and fish escape. Furthermore, sudden changes in water quality stressed fish and slowed their growth. The fish farmers secured their cages, monitored the cages and fish, moved the cages closer to the bank, and delayed or postponed the stocking date to prevent flood losses. These methods are comparable to those used by fish farmers raising tilapia in the Ping and Nan rivers. During the high-flow season, moving the cages closer to the bank can decrease the impact of floods. High flow and floods can cause cages to be destroyed, and the fish will use a lot of energy swimming against the turbulent flow. Cold weather has a tendency to reduce growth. Fish will consume less or cease eating in tropical locations when temperatures are below 25 °C due to low metabolism and excessive stress. This will have an impact on the rate at which disease develops. To cut losses, fish farmers kept a constant eye on their fish, postponed or delayed stocking dates or the entire crop over the winter, and harvested fish ahead of schedule. Changing the stocking date, lowering the stocking density, reduced fish growth during weather variability, and paying attention to weather forecast news were among them(Lebel, Whangchai, Chitmanat, & Lebel, 2015).

Climate change adds to the numerous stresses that fish stocks already face like overfishing, habitat loss, pollution, disturbance, and invasive species. Other anthropogenic influences, which frequently have a much higher and more immediate effect, must be considered when evaluating the effects of climate change. Climate change is one factor that can influence the climate. Variations in solar radiation, deviations in the Earth's orbit, mountain-building and continental drift, and changes in greenhouse gas concentrations are all examples of these processes. Some elements of the climate system, such as the seas and ice caps, respond slowly to climate change. As a result, it may take centuries or more for the climate system to fully adjust to new external changes. Changes in baseline socioeconomic conditions, adaptability, and biophysical processes were all assumed in several of the investigations. Climate change is a shift in the statistical distribution of weather over time scales ranging from decades to millions of years. It could be a shift in the average weather or in the distribution of weather occurrences around that average. Climate change may be localized to a single location or affect the entire planet. Anthropogenic climate change, also known as global warming or anthropogenic global warming, is a type of climate change caused by humans. Direct effects alter growth, reproductive capacity, mortality, and distribution by affecting physiology and behavior. Many people's food and water security will be impacted by climate change impacts such as more frequent and severe floods and droughts.

9.5 GREENHOUSE GASES

CO_2 and greenhouse gases (Figure 9.4) in the atmosphere are changing several aspects of the Earth's climate, oceans, coasts, and freshwater ecosystems, including air and sea surface temperatures, rainfall, sea level, ocean acidity, wind patterns, and the intensity of tropical cyclones, all of which have an impact on fisheries and aquaculture. Fishermen, fish growers, and coastal residents will stand the loss of these effects, with fewer secure livelihoods, fluctuations in the availability and quality of fish for food, and increased health, safety, and property threats.

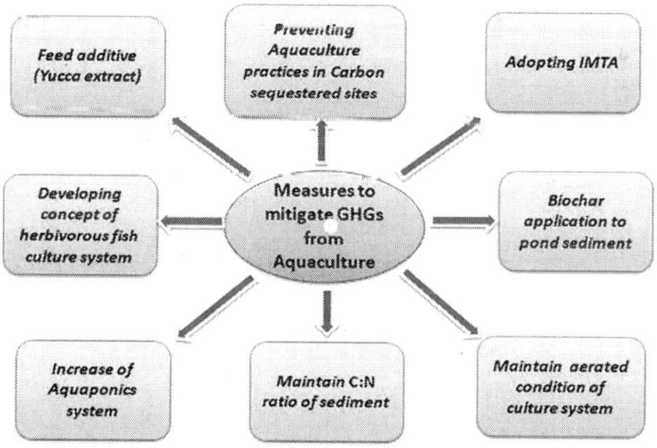

Figure 9.4: Greenhouse gases.

Source:https://www.researchgate.net/profile/Chittaranjan-Raul/publication/344807131/figure/fig2/AS:949366860832768@1603358106053/Mitigation-strategies-for-greenhouse-gas-emissions-from-aquaculture_W640.jpg

Poverty, a lack of social services, and necessary infrastructure have already made many fisheries-dependent communities unstable and vulnerable. Overexploitation of fisheries resources and deteriorated habitats exacerbate the vulnerability of these populations. Climate change has significant ramifications for food security and livelihoods in small island states and many poor countries. Climate change is impacting biological processes and altering food webs by changing the distribution and productivity of marine and freshwater species. The effects for aquatic ecosystems, fisheries, and aquaculture, as well as the people who rely on them, remain unknown. Rising ocean temperatures and acidification indicate that the ocean carbon sink's capacity would deteriorate over time, raising global concerns articulated in the Monaco Declaration. The increase in CO_2 levels owing to emissions from fossil fuel combustion is the most concerning of these anthropogenic effects, followed by aerosols and particulate matter in the atmosphere, and cement manufacturing. In order to forecast future climate impacts on marine ecosystems and fisheries, researchers must first understand how climate change may alter decadal and shorter time scale variability. Climate change has an impact on glaciers and the sea level(Navarro, Alves, & Ribeiro, 2021).

Climate change puts coastal and fishing populations, as well as countries that rely on fisheries, at risk. Bangladeshi fishing towns are facing not only rising sea levels, but also increased flooding and typhoons. Fishing communities along the Mekong River produce approximately 1 million tonnes of fish each year, and saltwater intrusion from increasing sea levels and dams would harm livelihoods and fish productivity. Based on an ensemble of climate models, the best current forecasts of changes in global marine primary production show an increase of less than 10% by 2090, which appears to be a manageable change. The underlying models, on the other hand, are highly unknown, and there are likely to be significant regional variances. Changes in species abundance can have a number of economic and social consequences, both direct and indirect. Changes in the physical environment, like weather and ice conditions, can affect where and when fishing is physically possible, as well as fishing prices. The worth of the current crop in the short term, to the degree that changes in harvests are not offset by changes in prices. Tilapia can live in a variety of environments, from pure freshwater to hypersaline water. Waterholes in ephemeral rivers, reservoirs, lakes, ponds, agricultural dams, rivers, creeks, drains, swamps, salt lakes, and tidal areas have all been documented in Australia. They will live in a variety of environments. Tilapia also have a wide range of dietary tolerances and are exceedingly adaptable, so they may compete with native species in areas where their diets and habitats overlap. Tilapia have been observed to switch their diet from detritus, algae, macrophytes, and other organic materials to invertebrates and other fish depending on food availability. They can be completely herbivorous, omnivorous, or completely carnivorous, even cannibalism, and prey on native fish eggs and larvae.

At all stages of aquaculture, temperature is the master abiotic element that governs and limits fish development and physiology. Changes in fish growth, hemato-physiology, metabolism, immunological, and molecular stress responses to severe temperature occurrences are all influenced by major physiological restrictions at the individual level. The effects of extreme temperatures on fish differ between and within species, depending on the severity of the stress and other factors. Global temperature trends have shifted dramatically as a result of climate change, and are expected to continue to do so. According to temperature trend assessments, the total worldwide land surface area subjected to extreme cold and extreme heat waves is growing. Functional tissues and organ damage produce specific cellular enzymes into the bloodstream during temperature stress in fish and other animals, which can be employed as stress indicators. In Nile tilapia, *O. niloticus*, ALT, AST, and WBC levels were considerably elevated after 3 days of stress at 13°C from 25°C(Panase, Saenphet, & Saenphet, 2018). Overall, some findings imply that fast drops in water temperature can trigger stress responses and compromise hematological markers. Several studies have found that extreme temperature occurrences can disrupt fish hemato-biochemical parameters. Extreme temperature occurrences cause significant cellular damage and anomalies. Fish must balance a large amount of energy to repair this damage and recover cellular equilibrium, which has an impact on growth, immunity, and disease resistance. Aside from the complexity of interactions among biotic and abiotic factors, pre-exposure history, and compensatory adaptive mechanisms, stress response variation occurs across a wide range of species, and even within the same species, different strains, size, age, and location must be taken into account in future research.

9.6 TEMPERATURE

Temperature influences feed intake, digestion, absorption, and assimilation in ectotherms. To survive extreme temperature events, fish sacrifice extra energy in recovering homeostasis. Energy is proportionately allocated in fish under non-stressed situations to maintain storage, development, growth, activities, and body maintenance. Maintenance costs, on the other hand, rise during a stressful scenario to cover higher energy requirements in stress reduction and cellular damage repair, which counteract energy allotted for growth, development, immunity, storage, and reproduction. As a result, growth, reproduction, and immunological performance suffer. Temperature stress has been linked to growth suppression, alterations in reproduction performance, and sex ratio. Temperature rises within the aerobic scope are likely to be beneficial for some teleosts inside their chosen thermal window found that a slight temperature increase results in higher energy and metabolic activities, higher absorption and assimilation rates, and faster physiological activities lead to a higher growth rate. Laboratory-controlled temperature studies, on the other hand, showed that temperature could reach a point where it was detrimental to growth performance. Temperature stress is associated with decreased development, feed intake, and food conversion efficiency.

Changes in ambient water temperatures and other factors affect spawning, reproduction, phenology, sexual maturity, and sex differentiation by interfering with the steroidogenesis and gametogenesis processes. Temperature events caused by climate change are likely to impair teleost reproductive success. Through endocrine disruption, temperature fluctuations either enhance or retard gametogenesis and maturation. Increased temperature during gametogenesis in Atlantic salmon, *S. salar*, resulted in decreased steroid synthesis in the gonads, liver vitellogenin production, and changes in hepatic estrogen dynamics. Reduced reproductive investment and gamete viability are the end results of these modifications. High-temperature exposure impairs gonadal steroidogenesis, delays, or slows pre-ovulatory shift throughout several stages of gonadal maturation. During elevated temperature exposure, suppression of spermiation was found in Northern pike, Esox lucius, Atlantic salmon, *S. salar*, rainbow trout, *O. mykiss*, Climate change-induced temperatures significantly influenced sperm motility and quality in coruh trout (Figure 9.5). *Salmo coruhensis* (Figure 9.6), rainbow trout (Figure 9.7), *O. mykiss*, sturgeon, *A. brevirostrum*, and streaked prochilod.

Figure 9.5: Coruh trout.

Source:https://www.greatbarrierreefs.com.au/wp-content/uploads/10-779x1024.jpg

Figure 9.6: Salmo coruhensis.

Source: https://www.fishbase.se/images/species/Sacor_u6.jpg

Figure 9.7: Rainbow trout.

Source:https://www.wildtrout.org/imager/general/111/rainbow_trout_983c404359247ecee4033ec1e6bdab0c.jpg

To comprehend the consequences on aquaculture and fisheries output, it is critical to understand how fish will survive with future climate change-induced severe temperature occurrences, as well as other environmental difficulties. Extreme temperature occurrences can dramatically impact fish development, reproduction, metabolism, physiology, behavior, and survival. Although fish can survive with a certain amount of heat stress by increasing their growth and physiological fitness, there is little information on how they cope in the face of extreme temperatures. To comprehend future climate change-induced fish ecophysiology, more species-specific and environmentally realistic studies are required. To understand unexpected results under various climate scenarios, it is also required to select a larger range of stressors and model fish species. There is a need for more studies into non-lethal endpoints for tracing stress events in fish. There are no studies that use a collective strategy to look at fish stress responses, life cycles, feeding strategies, or behavioral responses under a variety of culture settings or regimes. Exploring these possibilities could pave the way for climate-smart aquaculture research and development. Furthermore, extreme temperature occurrences affecting fish in open aquaculture systems should be addressed. Stress-tolerant fish species and strains, alternate species farming, and relocation of aquaculture locations could all be alternatives for mitigating severe temperature events. To limit the harmful effects of extreme climatic occurrences, more applied study is needed to discover the most suitable alternative solutions. Dietary and nutritional mitigation techniques, in addition to these, could be a promising alternative for managing thermal stress. Nutritional measures as a mitigation approach to prevent climate change-induced thermal stress will require a wide range of studies with a variety of extra ingredients. Another essential issue to note is that the majority of climate change effect research has been conducted on aquaculture species from the United States and Europe. In Asia, the world's largest seafood source, there is only a little research available for a few flagship aquaculture species. To effectively prepare for global climate change, more focus should be paid to studies on Asia's vastly understudied aquaculture species.

Climate change will exacerbate the African coastal zone's current physical, ecological, and socioeconomic pressures. The advantages derived from the development of fisheries are substantial. Fisheries and aquaculture play vital roles in food supply, economic production, and nutrition at all levels, from local to global. Around 43.5 million individuals work directly in the fishery sector around the world, with the vast majority working in underdeveloped nations(Berg, Mulokozi, & Udikas, 2021). When you factor in those employed in related processing, marketing, distribution, and supply industries, the sector employs approximately 200 million people. In Sub-Saharan Africa, fisheries employ roughly 10 million people and are the main or sole source of animal protein for 20% of the population. As a result, the sector plays an important role in increasing food supply and reducing food security threats in various agrarian and food-insecure countries in the region. Fish is a commonly traded food in the region. Fish commerce aids economic growth by providing a valuable source of cash to settle foreign debt, fund national government activities, and import food for home consumption, all of which contribute to national food security and diet diversification. However, in national economic planning, the sector's benefits are frequently overlooked or overstated.

While the value of fisheries is frequently overlooked, the consequences of climate change for these industries, as well as coastal and riparian populations in general, are difficult to overlook. Climate change, in addition to numerous other concurrent stressors such as overfishing, habitat degradation, pollution, and the introduction of new species, poses a substantial threat to fisheries. Globally, 20% of targeted fishery resources are moderately utilized, 52% are completely exploited with no additional increases expected, 19% are overexploited, 8% are depleted, and 1% are recovering from past depletion, compared to the level that would strengthen maximum sustainable yield(Watanabe, Losordo, Fitzsimmons, & Hanley, 2002). Changes in the aquatic environment's biophysical properties, as well as the incidence of extreme events, will have a substantial impact on the ecosystems that support fish. This will have a variety of consequences for food security. For starters, the extinction of some fish species reduces the amount of fish available for local consumption. Second, the migration of many fish species to aquatic habitats with ideal climatic conditions would have a huge impact on fishermen who are unable to track fish owing to political and economic constraints. Small-scale fisheries supply the majority of fish gathered for export in many developing countries, reduced fish output would result in fewer profits from fish export, resulting in reduced capacity to import food and exacerbation of national food insecurity. Furthermore, as evidence of climate change's consequences on aquatic ecosystems grows, the resulting impacts on food security and fishery livelihoods are anticipated to be significant, but this remains a neglected topic in climate adaptation programs.

9.7 OCEANS

The oceans have a key role in global climate regulation. They absorb a huge amount of heat emitted globally since their heat capacity is 1000 times greater than that of the atmosphere. Changes in ocean temperatures have the potential to alter the dynamics of the region's aquatic habitats. Changes in ocean dynamics may cause changes in fish migration patterns and, as a result, a reduction in fish landings, particularly in coastal fisheries. Increased ocean temperature, for example, may alter upwelling along the Gulf of Guinea, making ocean waters unsuitable for fisheries, resulting in a decrease in and possibly collapse of fishing activity. Inland waters are extremely vulnerable to climate change and might be severely damaged. Increased sea surface temperature is one of the stressors that SSA's inland waters are facing. Warming in Africa will most likely be more than global annual mean warming throughout the continent and in all seasons, with dry subtropical regions warming more than wetter tropics, resulting in a decrease in rainfall. Biodiversity is also threatened by rising water temperatures. Fish, in general, have a preferred temperature that optimizes physiological activities. When water temperatures rise above a species' maximum acceptable threshold, the species' survival is jeopardized. The quantity of tadpoles and frogs has decreased dramatically as a result of warming rivers, which have boosted the number of predator fish in an area they have never previously inhabited.

Climate change has the potential to raise or decrease water salinity in a variety of ways. This shows that tropical oceans are more likely to be affected by rising water salinity than waters at higher latitudes. The salinity of some freshwater ecosystems is expected to rise. By impairing the organisms' ability to osmoregulate, such physical alterations will have a negative influence on the population of both plankton and larger prey fish species. Changes in salinity have a deleterious impact on zooplankton populations, according to several empirical investigations, especially in freshwater settings. Small changes in salinity have a negative impact on zooplankton communities in low-lying, coastal, tidal lakes, and wetlands. They also warn that fluctuations in zooplankton abundance could wreak ecosystems of these valuable but vulnerable ecosystems. Changes in zooplankton numbers or other planktonic primary and secondary producers interrupt the food chain, which has a significant detrimental influence on fisheries. Oceans are thought to be capable of absorbing the majority of anthropogenic CO_2 emissions. CO_2 is water-soluble and transforms to carbonic acid reversibly. The world's seas are acidifying at an alarming rate as a result of this chemical process. While this helps to delay global warming, increased acidity in seawater as a result of dissolved CO_2 has a deleterious influence on ocean organisms. The ecosystem's influence is difficult to measure because various species respond to pH changes differently at different phases of their lives. Despite the fact that the scientific base is currently inadequate, there is an obvious need for timely advice and a balanced perspective on the potential implications on marine fisheries. Assessing the direct and indirect chemical impacts on valued marine ecosystem services such as fisheries is required to quantify the implications of ocean acidification on human communities.

Changes in physiological processes such as calcified structure growth, otolith development, and fertilization success are all direct impacts. Lower growth and reproductive output, increased predation and mortality, changes in eating rates and behavior, reduced immunocompetence, and reduced heat tolerance could all result in direct effects on the complete organism. Changes in predator or prey abundance, effects on biogenic ecosystems such as coral reefs, and changes in nutrient recycling are all examples of indirect effects. Increased CO_2 levels in seas have the ability to narcotize male gametes, implying that acidification may hamper fertilization, increasing sperm shortages and having catastrophic consequences for marine life. The growth of plankton and invertebrates at the bottom of the food chain could be slowed by ocean acidification. Acidification can alter productivity at specific trophic levels, altering the complex food chain of aquatic ecosystems and having an impact on fisheries production. A decline in populations of calcifying species such as mollusks is one of the most likely socio-economic effects of ocean acidification in SSA. The link between future ocean primary production and climate change is projected to be a major limitation on fish and fisheries productivity. The availability of sufficient and acceptable food is regarded to be critical for fish larvae's survival during the planktonic stage. Climate-induced variations in the distribution and phenology of fish larvae and their prey can thereby affect recruitment and production of fish stocks, in addition to the effects of changes in productivity. The case of Lake Tanganyika is one of the few studies accessible. Lake Tanganyika has historically sustained one of the world's most productive pelagic fisheries, with an annual harvest of between 165,000 and 200,000 metric tonnes in previous years, accounting for 25–40% of animal protein supply for the surrounding countries' populations.

9.8 STABILITY

The stability of the water column has risen due to a rise in surface-water temperature combined with a regional decrease in wind velocity, limiting mixing and deep-water nutrient upwelling and diffusion into surface waters. A 20% loss in primary productivity translates to a roughly 30% fall in fish yields. The drop in catch was followed by the collapse of previously robust seasonal patterns in catch, implying a decoupling from ecological processes caused by decreasing hydrodynamic patterns. These alterations in the pelagic fishery are consistent with a shift in ecosystem functioning over the entire lake. Although there are many elements that influence primary production in aquatic environments, one of the most important is the increase in surface temperature. If air temperature rises by around 1.7 degrees, as expected over the next 80 years, thermal stability will increase further and productivity would decrease in these huge lakes. They also warn that the human consequences of such gradual environmental changes could be disastrous in this part of the world, where huge lakes are vital natural resources for regional economies.

In most SSA nations, small-scale or artisanal fisheries companies are thriving. They have the potential to be long-term investors. Fishermen have a strong chance of selling their catch at good rates at the local, regional, and national levels, while retaining improved fish production systems that address over-investments in fishing, excessive water consumption, and environmental deterioration. Mechanisms must also be suited to reduce operating expenses and fuel consumption, adopt less expensive technology, and have a smaller environmental impact, resulting in increased employment opportunities. These chances, however, are rarely taken advantage of. They are unable to invest in quality businesses that are long-term viable because of the low prices they receive. Because the fishermen are not properly organized, they do not buy in bulk or sell their products in bulk, which would lower manufacturing costs and raise sale prices. Small-scale fisheries and other private-sector organizations have essentially no coordination, leaving them in a weak competitive position.

Small-scale fishermen face similar difficulties in obtaining loans; loan amounts are frequently insufficient, and repayment schedules are not tailored to the cash flow of the firm. Because high credit worthiness or collateral is required to receive larger loans, many small enterprises must start with small loans. As a result, they rely on their meager savings to get by. Furthermore, repayment schedules are frequently not tailored to the cash flow of the business: repayments are made weekly, while most of these enterprises' revenue is seasonal. Many countries do not offer seasonal loans, and those that do are only for small amounts because the risk to lenders is higher. Artisanal fisheries also lack access to low-cost fishing technology that is ideal for a variety of commercial species, as well as the knowledge to operate them. Modernizing fishing practices and operations, for example, through better fishing boats, fishing gear, and post-harvest equipment, is one example of low-cost technology adoption. Competitiveness and income can be improved at a greater level by introducing fish processing technologies that boost the value of products.

Researchers are concerned about the effects of global climate change on fish populations and fisheries, but there has been little effort to incorporate observed changes or event-based thinking into management models and mainstream fisheries governance into adaptation programs. To efficiently manage fisheries resources, fisheries governance provides policy instruments and regulatory measures, as well as links governments with stakeholders such as communities, civil society, and the commercial sector. It's vital for determining if the negative effects of climate change on fisheries can be reduced and whether the industry can continue to deliver economic, livelihood, and food security advantages. Governments should facilitate food security for all households and individuals, not by physically delivering the needed food to all citizens, but by facilitating a social and economic environment that allows all citizens to ensure their own food security. The similar argument may be made for improving community resilience to cope with and adapt to climate change impacts, as well as promoting both economic and ecological sustainability of fishery resources. Adaptation is a dynamic social process in which civilizations' ability to adapt is influenced by their ability to act collectively. Co-management or cooperative management in the aquatic environment, in which a variety of stakeholders share responsibility for decision-making and administration of marine resources, can help local communities create resilience to climate change. As a result, governments must invest in strengthening and utilizing the social capital of fishing communities, which is the poor's most valuable asset. This could help fishers participate in the development of adaptation policies and successfully manage their resources by actively participating in the implementation and monitoring of regulatory measures aimed at enhancing their livelihood. Societies with a high reliance on fisheries for animal protein are more likely to be poor and malnourished. Efforts must be made to diversify fisher communities' sources of income and minimize their reliance on the sector.

Many people have advocated for the use of economic incentives, primarily as a realistic way to internalize environmental costs where regulatory approaches have proven ineffective. Payments for ecosystem services (PES) are a type of economic incentive mechanism that rewards resource users for better natural resource management practices or compensates them for the advantages lost as a result of complying with specific natural resource management regimes. The approach has been used in a variety of terrestrial environments, with a focus on forest ecosystem services. Attempts to apply the same technique to marine and coastal ecosystems have recently been made. In contrast to terrestrial environments, implementing incentive mechanisms in water ecosystems can be difficult.

Aquatic resources and ecosystems are mobile, and aquatic environment ownership and property rights are frequently ill-defined and unrecognized, making implementation, monitoring, and enforcement extremely difficult.

Nonetheless, by assisting fisher communities in diversifying their sources of income and providing better community services such as post-harvest technology development and value addition, these mechanisms can play a significant role in incentivizing them to co-manage and restore their fisheries resources. At the community level, co-management refers to the participation of fishermen in the management of their fisheries in order to enhance their livelihoods. However, existing co-management arrangements have generally focused on fish resource conservation and management, rather than using them to assist economic development in fishing communities or as a tool for poverty reduction. As a result, the effectiveness of the fisheries co-management strategy has been questioned in numerous instances. Economic incentives for fishing villages may be able to solve the situation. Economic incentives could compensate fishing communities for the short-term loss of advantages from restricted fishing. This could provide livelihood benefits in addition to motivating communities to conserve their fishing resources, thereby addressing the issue of food insecurity.

CHAPTER 10

BENEFITS OF TILAPIA POLYCULTURE

CONTENTS

10.1 INTRODUCTION

By 2050, the population of the world is predicted to increase from 6.1 billion to 9 billion people. As a result, demand for fish-based foods is expanding. From its current output level of 101 million metric tonnes, the demand for aquatic goods for human consumption will expand to 121.1 million metric tonnes by 2010. This extends beyond the supply of total catch fisheries. Aquaculture will mostly make up for the shortage in supply. Aquaculture has been the world's fastest-growing animal production sector since 1970, growing at a rate of 9.2 percent per year on average, compared to only 1.4 percent for catch fisheries(M. Wang & Lu, 2016). Aquaculture, fish trading, fish processing, and fishing gear manufacture employ a large number of people in primary, secondary, and tertiary fishery-related sectors. There is room for improvement in Bangladesh, since the majority of fish farmers are poor and use old or rudimentary methods. Improved livelihoods, including nutrition, food security, and income, require novel and easily improved technologies that are accessible and relevant to disadvantaged farmers.

10.2 SOURCES

The principal food and nutrient sources in vast and semi-intensive aquaculture ponds are in situ generated algae and imported organic matter. Algae use solar energy, carbon dioxide, and inorganic nutrients to make organic matter that can be eaten directly by culture animals, zooplankton, benthos, invertebrates, and other microbes, or degraded by bacteria, fungus, and other microbes. Indirectly, all species that contribute to the heterotrophic food chain, as well as bits and pieces of decomposing organic materials, help to feed culture animals. The phytoplankton population is thought to be the most essential in terms of energy fixation for feeding the food chain, especially in aquaculture. However, especially in shallow waters, macrophytes and periphyton represent a significant and frequently dominant contributor to primary production. Because grazing on biofilms is more efficient than filtering planktonic algae, periphyton-based ponds will achieve a more efficient energy transfer than typical ponds with herbivorous or omnivorous species culture. Furthermore, periphyton communities are more stable than phytoplankton colonies, which can readily collapse in severely eutrophic ponds, depleting oxygen. Periphyton-based ponds are easier to maintain than regular ponds.

Although tilapia originated in Africa, they are now found in most Asian and Latin American countries. Despite the fact that there are over 200 species of tilapia accessible for culture in different parts of the world, the Nile tilapia (*Oreochromis niloticus*) became a focus species. Over the last five decades, freshwater fish culture has progressed. The rise of aquaculture has been aided by a reduction in catch fisheries. The common carp (*Cyprinus carpio*) was the first species introduced for aquaculture in most nations, but various additional species have since been introduced, including a variety of carp, tilapia, catfish, freshwater prawn, and others. People began to culture many species in the same pond to reduce risk and to have fish of different flavors. However, as they learned that different species have different feeding habits, they began to select fish that are compatible or will not compete in terms of space and food. Polyculture is the practice of growing two or more species in the same system. The major goal is to boost fish production by utilizing natural feeds like phytoplankton, zooplankton, detritus, grasses, and macrophytes available in the system. This is feasible because tilapia only eat phytoplankton and/or detritus, which are the two types of food they prefer. Because there are so many different fish in the tank, any food that the tilapia doesn't consume will be eaten by other species and won't go to waste. Adding one or a few more kinds of fish would not increase the farmer's costs, but would enhance the production, resulting in a higher net profit when the fish are sold.

Figure 10.1: Tilapia polyculture.

Source: https://www.globalseafood.org/wp-content/uploads/2001/06/Fitzsimmonspic1.jpg

10.3 POLYCULTURE

Polyculture is typically done in ponds (Figure 10.1). The type of fish stocked and the amount of fish stocked are totally up to the farmer's preferences and experience, but the feeding habits of the various fish species, as well as whether there is a market for them, must be considered. Grass carp, for example, would be really beneficial because it eats grass and cleans up pond dykes. Grasses grown on the dyke can potentially be used as a source of feed. It is regarded as the most important species for polyculture in China because it eats a lot of grasses and excretes green water that nourishes other plants. More importantly, it can be sold to Chinese populations at premium prices. Chinese farmers primarily raise tilapia, with silver carp serving as a cleaner fish that can filter plankton. Another reason is that silver carp die first when water quality deteriorates, which acts as an indicator. In Thailand, the silver barb (*Puntius gonionotus*) serves as an indicator since it is susceptible to death if water quality deteriorates.

Figure 10.2: African catfish.

Source: *https://images.thefishsite.com/fish/articles/africa/african-catfish.jpeg?profile=article-inline@maximum*

10.4 AFRICAN CATFISH

African catfish (Figure 10.2) is effective at reducing tilapia recruitment or unwanted fry, particularly in mixed-sex tilapia production systems. They're also good for monosex culture because a few females can still produce fry. If brooders are controlled in hapas, catfish are purposefully stocked to limit tilapia recruits in broodstock ponds. Whatever type and number of fish are stocked, keep in mind that they should all reach market size in the same amount of time, such as 6–7 months for tilapia. Nile tilapia reproduce without the use of hormone injections in culture systems, which is advantageous in some rural areas where fish farming is primarily for family consumption(Ibrahim & Naggar, 2010). Small fry is generated as a by-product and can be sold as fish seed or used to fill other ponds, providing farmers with an extra source of revenue. However, because about a quarter of the fish are undersized and unsaleable, high reproduction has been a major disadvantage of this species for commercial farming. The main issue with fingerlings in grow-out ponds is that they compete for feed and space with adult tilapia, causing growth to be hampered. Monosex tilapia fingerlings appeal to farmers for this reason. However, in many regions, particularly rural areas, producing monosex fry is not viable, hence most farmers in these areas use carnivorous fish species to restrict tilapia recruits. Rearing tilapia alongside a predator fish can minimize the number of recruits. Polyculture is typically semi-intensive in nature, with numerous harvests rather than a single final crop. Farmers can increase their total fish yield while also increasing their income. They can also harvest whenever there are favorable market prices for certain species or whenever they need to fulfill family or community rites, which could be multiple times a year. Because the farmer must empty and dry the pond before filling, fertilizing, and stocking the fry for the second crop, there will be a gap between the two harvests. When compared to a single harvest procedure, the overall volume of fish produced can be much higher. Polyculture pond farmers typically harvest many times over the course of several years. Assuming a two-year comparison with monoculture, the total number of fish harvests might be five or six, depending on the species stocked and management approach. Some harvests may be of one species, while others may be of a different species.

Farmers can enjoy diverse species if they consume at home, as a consequence of maximum use of available food in the culture system, which results in the largest production per unit of culture system. There's a lower chance of market price fluctuations. Even though one species' price is low, another species' price may still be high. Fish eat natural food from various regions of the culture system, such as the bottom, column, and surface, resulting in low feeding expenses. It's a lot more intricate and tough to deal with. Farmers must be knowledgeable about the cultivation procedures of multiple fish species. Farmers must worry about marketing a variety of species, and it might be difficult to obtain a fair price for all of them. Because it necessitates additional products, facilities, and skilled personnel, the total cost of management and operation is significant. Farmers have traditionally grown cereal crops, vegetables, fruits, and spices, as well as raising animals such as buffalo, cattle, fowl, pigs, goats, sheep, and other livestock, all while living on the same plot of land. They have been using wastes from one component to feed another, whether consciously or inadvertently, in order to avoid having to bring in inputs from outside. Because nutrients are recycled within the system, these traditional methods are naturally sustainable, and their worth has just lately been recognized.

It has been customary in most nations for a farmer to grow crops, vegetables, fruits, and keep animals in the same area. Crop by-products, such as rice bran, wheat bran, and other grains, can be utilized to feed cattle, with a small portion going to the fish pond. Farm by-products are reused in an integrated system. Livestock consumes the majority of inputs and generates a significant amount of manure, which was previously utilized mostly for crops, vegetables, and fruits. Farmers have lately discovered that using it as a pond input to produce fish, which is more effective in terms of nutrient use and produces a better income, is a viable option. This demonstrates that farmers require extremely few external inputs. If a significant number of external inputs must be purchased, this may not be long-term viable. Farmers can maximize their produce from all aspects of agriculture, including cattle, fish, crops, fruits, and vegetables, by practicing integrated farming.

10.5 FARMING

Farmers may consume a large portion of their food, providing nutrition to their families. Farmers, on the other hand, are becoming more commercial, and they may now trade their produce and acquire food supplies as needed. They frequently raise high-value crops and buy lower-cost ones for household consumption. The majority of tilapia farming takes place in ponds, which are fertilized with feed, dung, or inorganic fertilizers. Natural food species provide significant amounts of nutrients to the fish in these settings. It was discovered that natural foods contributed 50% to 70% of the growth of tilapia polyculture with carps in ponds receiving supplemental feeds, based on stable carbon isotope studies of the fish and the food sources. In the intensively fed pond cultures, stomach analyses revealed that up to 50% of the tilapia stomach contents are natural food, showing that natural pond productivity supplied a significant amount of nutrients. The amount of this contribution to the fish's nutrient needs for maximal growth will be determined by the pond's production and fish density.

The polyculture system is used by the majority of pond-based tilapia farmers in Bangladesh, China, Taiwan, Thailand, and Vietnam, whereas the monoculture system is used by the majority of tilapia farmers in the Philippines. The farming practices used in these countries differ depending on the quality of the farmland and the farmers' willingness to invest. In Bangladesh, for example, most farmers do not utilize commercial feed, while in Vietnam, farmers only use a tiny amount of commercial feed. On the other hand, most tilapia growers in freshwater ponds in China, Taiwan, Thailand, and the Philippines are also widely integrated with agriculture and animal farming in Southeast Asia; particularly in Indonesia, Thailand, Vietnam, Cambodia, and Myanmar. Since the 1980s, integrated livestock-fish systems have been a widespread practice in Thailand's Central area, particularly on relatively large farms.

In Israel, sex-reversed male *O. niloticus* and *O. aureus* hybrids are polycultured with carp in earthen ponds or monocultured with high stocking density in plastic-lined ponds. Earthen pond aquaculture is the most common form of aquaculture system in Egypt, where only wastelands are permitted to be utilized for fish due to their high salt and alkali content and poor drainage. Semi-intensive aquaculture, which is largely done in ponds, accounts for around 75% of the country's total aquaculture production, with the majority of farms located in the Nile Delta's northern and eastern regions. Earthen ponds are the most important small-scale, monoculture of tilapias at the household level in Africa, accounting for 38-93 percent of overall tilapia production. Productivity ranges from 0.5 metric tonnes per hectare per year in large small-scale fishponds to 16 metric tons per hectare per year in commercial ponds. *O. niloticus* is the most common species employed(J.-Q. Wang, Li, Dong, Wang, & Tian, 1998). Other species, such as *T. zillii* and *O. rendalli*, are also cultivated with *O. niloticus*. Tilapia pond production on a small scale is frequently combined with other agricultural endeavors such as vegetables, rice, and other field crops. Tilapia farming on ponds on a big scale and on a semi-intensive basis is also practiced in some African countries, such as Zambia and Cote d'Ivoire.

These are mostly produced in large-scale systems that require just basic equipment and minimal care. However, rising demand for higher-trophic level species like sea bass, salmonids, some catfish, and shrimp from rapidly expanding middle classes and urbanization is likely to lead to a shift toward more intensive, high-tech farming, including recirculating aquaculture systems. There's also a push to improve aquaculture's environmental performance by developing systems like multi trophic aquaculture, which uses low-trophic-level species to utilize the nutrient-rich by-products from high-trophic-level species; for example, farming shellfish in fish farm drainage canals, polyculture, and rice/fish culture. This could include focusing on species with a proven production base and value chain, with a preference for local, indigenous species if available. For example, if widespread low-trophic species cultivation already exists, this should be used as a starting point. It's also critical to recognize and learn from efficient systems such as polyculture, as well as unpleasant experiences and events, such as earlier disease problems. Another facet of this is a general philosophy of avoiding huge, costly infrastructure projects, particularly those with long-term maintenance and support requirements. If infrastructure development projects such as hatcheries are being considered, they should be demand-driven, and existing facilities should be renovated rather than new builds wherever practical.

Because aquaculture is quickly developing in most countries without donor assistance, rather than fostering new output, one feasible strategy is to incorporate economic, social, and environmental sustainability elements into existing activity. Capacity building, diversifying output with new species or fostering polyculture, bettering company management and planning, or focusing on sectoral level governance are all possibilities. Another method along these lines is to add value to production along the value chain, either by improved production techniques that allow eco-labeling/certification, or through post-harvest processing, product creation, and quality assurance. In Bangladesh, average annual tilapia consumption was around 2.5 kg/person in selected inland areas, but there was no data on average per capita consumption across the country. The importance of farmed tilapia in satisfying the needs of disadvantaged people, as well as its demand in formal markets, is becoming clearer. Because there were few tilapias in markets in north-eastern Bangladesh, for example, there was a notion that they were unimportant. Many small-scale farmers have been discovered stocking tilapia in polyculture with carps in small ditches and borrow pits close to their homes. Tilapia is the most common fish consumed by the poorest households. Indian and Chinese carp polycultures, as well as minor indigenous species, accounted for 85 percent of overall aquaculture production. Monocultures, primarily of catfish and tilapia, accounted for barely 3% of overall aquaculture output. There are currently no accurate figures on tilapia production. Small-scale tilapia farming, on the other hand, is probably more common than statistics show. Fish harvests featuring tilapia from small ponds and rice fields are frequently used primarily for home consumption, particularly in the country's poorest regions, such as the northwest. Because of its success in other Asian countries and rising market acceptance, tilapia aquaculture is gaining popularity. In addition to BFRI's and the government's initiatives, private sector hatcheries, such as GIFT, are boosting the supply of high-quality tilapia seed.

10.6 TILAPIA

Tilapia has a lot of promise as an alternative and additional species of farmed fish in Bangladesh, both monoculture and polyculture. If tiny homestead fishponds are included, the total number of fishponds in the country is likely to approach 5 million. This demonstrates the importance of fishponds in rural communities and households' livelihoods, nutrition, and social fabric. Bangladeshis consume 60–80 percent of their animal protein in the form of fish. Inland fish consumption per capita was greater than the national average in 1999. In selected inland areas in 1998–1999, average tilapia consumption was around 2.5 kg per capita, which was lower than silver carp and rohu intake. The yearly per capita tilapia consumption was 2.2 kg for rural fish producers and 2.6 kg for both rural non-producers and urban consumers in selected inland locations. If GIFT is successful in increasing the supply of farmed tilapia in Bangladesh, this will help to meet the growing demand for fish. Because of institutional, technological, and economical obstacles, tilapia farming has yet to make a significant contribution to Bangladesh's freshwater aquaculture productivity. Shortages of village extension personnel, improper extension material, a consistent supply of high-quality tilapia seed, and limited access to financial capital are among the obstacles. Carp polyculture has been prioritized by the government over other farmed fish. Although supplies from private sector hatcheries are expected to increase, GIFT seed is not yet readily available. However, without GIFT, interest in tilapia farming and overall support for aquaculture genetics would have stayed lower than it is now. Because tilapia provides an economical option for fish eaters, new market prospects for GIFT seed producers and tilapia farmers are projected to grow. Tilapia seed is generated in hatcheries and nurseries from broodstock reared in ponds, fine mesh cages, rice fields, and other small water bodies. Tilapia is farmed as a monoculture or in polyculture with native and alien carps and other species. The majority of tilapia farming takes place in Thailand's irrigated Eastern, Western, and Central regions. For the local and export markets, some significant private firms are producing sex-reversed red and Nile tilapia, with plans to continue tilapia production in the next five years. Bangladesh is one of the world's most favorable countries for aquaculture. The combination of a large amount of shallow water with a sub-tropical climate creates ideal conditions.

In Bangladesh, tilapia is widely acknowledged as a suitable species for freshwater aquaculture systems. Tilapia culture lasts a few months, from April/May to November/December, to allow for multiple culture cycles. Tilapia culture is primarily seasonal, beginning before the summer and continuing until the winter. Culture season, stocking density, and culture duration are all critical elements in tilapia farming because hatchery-produced fry is used exclusively. In inland water bodies, tilapia can be farmed in small-scale, large-scale, low-input, or high-input operations. Polyculture, monoculture, and integrated farming are the three types of tilapia farming. Tilapia culture can be classified as extensive, semi-intensive, or intensive according to the amount of input. In rural Bangladesh, tilapia pond polyculture with other species has long been done. Polyculture is frequently done in extensive or semi-intensive culture systems with a low or medium amount of input. Polyculture is centered on making the best use of a pond's ecological niches in order to maximize productivity per unit area. Tilapia eat organic particles on the pond floor that are re-suspended by carp species in the polyculture system's sediment tabulation. In this way, tilapia can prevent the organic load in the sediment from being lifted, resulting in anaerobic conditions. Depending on the culture species, natural food availability, and supplementary diet, the stocking density of tilapia in a polyculture system may vary.

Shrimp is a key species in coastal Bangladesh's brackish water aquaculture, which is one of the country's most profitable industries. Shrimp production has recently been hampered by environmental degradation caused by climate change and disease outbreaks. Farmers have been looking for an alternative strategy where crop diversification, such as polyculture rather than shrimp monoculture, might play an effective role in sustainable aquaculture to overcome this predicament. Tilapia can compensate for a portion of the economic loss caused by unanticipated shrimp mortality in polyculture. Although tilapia is a common freshwater aquaculture species in Bangladesh, it may also be grown in brackish water systems as a diverse crop. Because they utilize distinct ecological niches, tilapia can supplement shrimp output due to cumulative environmentally favorable polyculture settings. Tilapia, for example, feed on the upper water layer, filtering phytoplankton and zooplankton, whereas shrimp prefer to graze on the pond bottom for substrate and debris. Furthermore, tilapia can withstand a variety of environmental stresses, including salinity stress ranging from 0 to 25 ppt. Bangladesh's tilapia production was first unsatisfactory. The productivity of tilapia has increased significantly as culture technologies have progressed, converting extensive to semi-intensive or intensive culture using monoculture rather than polyculture. In most culture systems, tilapia production has gradually grown. Tilapia production, on the other hand, varies based on the culture system and species. Pond systems under freshwater aquaculture had the highest average tilapia productivity, followed by seasonal water bodies and cage–pen culture. Tilapia develops to a marketable size in a short culture time, allowing growers to produce many crops per year.

Furthermore, polyculture with other aquatic species makes adopting specialized tilapia harvesting techniques more challenging. As a result, small-scale farmers in various places may use a variety of simple harvesting techniques. The harvesting methods used are determined by the size of the pond, the culture systems used, and the level of technology used. Small-scale tilapia growers, for example, in many countries of Africa and Asia, find it difficult to get harvesting nets and other equipment necessary for comprehensive pond harvesting. To address this issue, several farmers use partial harvesting techniques, which make use of locally accessible equipment. Because they favor batch harvesting, large-scale tilapia producers use more modern harvesting instruments like winches. To decrease stress and comply with animal welfare rules, they sometimes anesthetize the fish before harvesting. Partially harvesting is frequently done to remove huge fish and make room for lesser fish to flourish. The annual output of all-male tilapia grown in polyculture with juvenile freshwater shrimp (*Macrobrachium rosenbergii*) and partially harvested monthly was similar to that of a batch harvest technique. Partial harvesting has a lot of potential for small farmers that need to supply markets on a regular basis. This approach makes it easier to meet the demand for fish fry because not all of the fish for stocking are needed at the same time. Again, technological, biological, economic, and sociological issues must be considered while analyzing tilapia harvesting methods. Fish harvesting methods are one of the primary impediments to the growth of rural aquaculture in many rural locations. Another example of the usefulness of system analysis is the economics and profitability of artisanal and commercial tilapia production in Malawi. Polyculture of Nile tilapia and African catfish (*Clarias gariepinus*) in earthen ponds was found to be more cost-effective than other culture systems when considering local support services such as fry and fingerling supply and local market development.

In the United States, a computer simulation model was used to evaluate the economic and engineering aspects of tilapia production in a small recirculating culture system. Sensitivity analysis revealed that of all the operational factors studied, lowering feed costs and increasing FCR were the most effective in lowering production costs. Increased system profitability was achieved by increasing output capacity or lowering investment costs. Feed, processing and distribution, and sex reversed fry were the biggest variable expenses on a commercial scale, seawater pond production of Florida red tilapia in Puerto Rico, while salaries and depreciation were the highest fixed costs. The use of locally prepared feed, as well as the integration of hatchery and grow-out operations, is recommended for lowering operational costs and boosting return.

We can compare tilapia and rohu carp polyculture to the other systems because they are the two most commonly farmed fish among the farmers. This contradicts the assumption that fish harvest in the global South is mostly exported to affluent countries or intended for the country's wealthier population, as is commonly observed in the early stages of aquaculture growth. Fish farming tends to concentrate on high-value species at this stage, such as Indian big carps, of which rohu is the most popular. Rohu carp cultivation has a high consumer demand and expansion potential, making it one of the most significant farmed fish species in India and neighboring countries including Bangladesh, Pakistan, and Myanmar. On the other hand, we need to observe that low-value species are rising in popularity in Bangladesh, with tilapia being consumed more than rohu carp. Tilapia was first brought to poor countries, where it was farmed on a subsistence level to supply protein requirements. After carps, tilapia has risen to become the second most significant farmed fish, with Indonesia, China, and Egypt as the leading producers. The costs and revenue in the rohu polyculture group, for example, refer to the entire costs and revenue from the polyculture pond containing rohu carp, not simply the costs and revenue from rohu carp alone, because it was not possible to split the input use by fish type in a polyculture

pond. All financial information, such as costs and revenue, was provided in the local currency. We then translated them to US dollars using the World Bank's 2019 average exchange rate. Polyculture ponds account for 66% of the ponds, with a median pond depth of 1.5 meters.

Despite the fact that tilapia is typically produced as a monoculture, tilapia aquaculture has also been embraced as a polyculture with appropriate indigenous fish species in order to continue increasing production and improving sustainability. Tilapia has also been used in shrimp polyculture, offering an extra revenue crop while simultaneously lowering the prevalence and severity of viral and bacterial diseases in shrimp. Farmers have introduced tilapia as an extra polyculture species to traditional carp ponds in pond culture. In the states of Andhra Pradesh, Gujarat, and Tamil Nadu, tilapia polyculture with carps is possible. Polyculture of shrimp and carp accounts for 3-5 percent of total tilapia production in the United States.

10.7 STOCKING

Stocking programs with fast-growing Chinese and Indian big carps and tilapias were tried in both perennial and seasonal systems during the 1980s. Unprecedented drought periods, poor management, monitoring, and assessment, and limited seed availability were all blamed for the program's poor long-term performance. The government's ban finally put an end to the program, although this resource has yet to be fully utilized. It was fueled by the increased demand that came with the pace of Dry Zone recolonization after the country gained independence in 1948. In Sri Lanka, *Oreochromis mossambicus* became known as the wonder fish. The ability of tilapias to self-recruit and exploit a wide range of low-level feed niches, which are normally only exploitable by using polycultures of other culturable species, as well as their good eating qualities, are the main reasons for their success in meeting the needs of the poor, especially when used in extensive fisheries enhancements. *O. mossambicus*, which is well suited to Sri Lanka's shallow and highly productive perennial reservoirs, filled this artificial lacustrine niche in a way that none of the slower-growing indigenous riverine species could. Sri Lanka presents a unique case in a worldwide spectrum of acceptance for tilapiine enhanced fisheries due to its unique structure of water resources and history of aquatic resource production, but one from which useful regional lessons can be learned.

Although comprehensive trials based on the suggestions of an FAO/UNDP mission aimed at producing a national development strategy began

in the early 1980s, the potential of the seasonal tank resource for increasing inland fisheries production was recognized as early as the 1960s. Tilapia and milkfish were among the other species used in the stocking trials. The poor survival of fingerlings was linked to predation, with slow moving carps being particularly vulnerable. This is especially true of common carp, whose proclivity for browsing at the edges increases its vulnerability. The overall outcome of poorly managed tanks is the development of a small number of fish with a high average weight. Grass carp performance was also hampered by a lack of submerged vegetation. Milkfish or *Labeo dussemiri* and *Labeo dussemiri* or *Labeo dussemiri* had poor survival and growth rates. Bighead carp had the best performance of all the species used, yielding nearly 1 ton/ha in composite polyculture yields. Average harvest weights rarely fell below 0.5kg, and in some cases reached as high as 2kg. In other cases, tilapia yields were higher, but bigheads were of more marketable size, leading to higher reported profits.

In the absence of repeatable experiments of various polyculture combinations in the Sri Lankan seasonal tank context, it is recommended that equal numbers of suitable species with complementary, non-overlapping eating patterns be stocked. Bighead and catla, as well as mrigal and tilapia, should not be combined. Catla and silver carp should be stocked in a 2:1 ratio. Because of the short growth seasons and receding water spreads, stocking densities at full supply level (FSL) or conventional stocking densities based on a 50 percent water spread as a rule of thumb may be kept low. Despite losses to predation, it was observed that there is a poor association between stocking density and output, with stocking densities of less than 1,000 /ha^{-1} and 2,500 /ha^{-1} yielding equivalent yields of around 0.5 mt/ha^{-1} in several cases. This value should be increased as aquaculture gets more intensive with the addition of extra feeds and fertilizers. According to the researchers, the tendency for tilapia to switch from somatic to gonadal growth, resulting in stunting, is mostly a consequence of stocking density, which makes them unsuitable for seasonal reservoirs. Under equal stocking conditions, *O. mossambicus* will stunt at a smaller size than *O. niloticus*. However, catches of *O. mossambicus* that had entered naturally through nearby rivers ranged from 0.4 to 408 kg ha^{-1} in the reservoirs utilized for carp culture, indicating good potential. The link between stocking levels, water spread, and harvesting policies must be carefully considered when tilapias are stocked in limited water bodies. Early maturity is influenced by photoperiod and light intensity, and *O. mossambicus* females can spawn while they are just 5 cm long. Precious breeding in low-water and low-

oxygen environments could have a deleterious impact on other species in polycultures. In such cases, it's a good idea to collect larger individuals on a regular basis. The size and density of nest sites can be utilized as a measure of population health.

Even though tilapias were not included in the stocking scheme, they contributed significantly to the catch in two of the tanks. The considerable diversity in stocking procedures contributes to the variation in total yield. However, the total productivity and contribution of carp species were both high in the majority of the tanks, indicating a favorable return on investment. Because of their relative ease of production, only common carp and tilapia are anticipated to be the most extensively used species in community fingerling production programs. Although juvenile tilapia has the unique ability to consume blue-green algae, the species may transition between phytophagy, zoophagy, and detriphagy, resulting in some overlap with other specialized' feeding niches other than macrophyte feeders. Polyculture, on the other hand, is expected to result in a net gain in productivity over single populations of carp or tilapia. In this case, the indigenous *Labio dussemiri* makes a significant contribution. Smaller farmers selling to exporters typically keep only a few species in stock, with monoculture or simple polycultures of the more popular, resilient species being the norm. The capacity to market produces at a fair price is determined by the size of the enterprise and its closeness to Colombo, from whence all exports are air transported. Smaller fish growers that live outside of Colombo must carry their catch individually. Fish are usually transported in oxygenated pouches that can hold them for up to 10 hours on public transportation. These commercial issues are the primary impediments to the poorest rural people adopting decorative cultivation. Early attempts by development organizations to encourage women's groups to participate in pond-based ornamental production such as the Small Fishers Federation and the Anuradhapura Dry Zone Development Program failed because suppliers were dissatisfied with the low return on the number of fries provided, the unreliable timing of harvests, packaging issues, and the potential of these groups to culture only the simplest and most common varieties when mixed consignments(Wijenayake et al., 2005).

CHAPTER 11

GLOBAL IMPACT OF TILAPIA CULTURE

CONTENTS

11.1 INTRODUCTION

Although introduced species may provide significant benefits in some cases, the spread of aquatic species as a result of globalization has often had unexpected and harmful impacts on receiving ecosystems, human communities, and economies. Many freshwater fish species have relatively few negative impacts, especially when compared to the economic value of their production, according to global databases of introductions. However, different authors have come to different conclusions about the extent of negative ecosystem impacts compared to the advantages of aquaculture and fisheries, particularly those based on tilapia. Since 2002, tilapia have been bred in the United States. Tilapia fisheries have been one of the fastest-growing worldwide aquaculture industries, expanding twice as fast as salmonid or carp fisheries. Because of the fast spread of tilapia aquaculture, various decisions are being taken to introduce fish into new watersheds, which might have far-reaching ecological, social, and economic consequences as ecosystem services alter. The goods and other resources that directly contribute to human welfare through markets or other non-market values are referred to as ecosystem services. Changes in an ecosystem's biotic components, such as those produced by the introduction of new species, can boost some services while decreasing others. Increased tilapia production may result in a decrease in harvestable native species populations, habitat loss, or recreational and cultural ecosystem services associated with native species affected by tilapia(Chatvijitkul, Boyd, Davis, & McNevin, 2017).

Divergent viewpoints on the types of ecosystem services offered or impacted by tilapia preclude the development of a comprehensive policy framework for the species' management. The majority of reported freshwater fish capture production comes from feral tilapia in some countries: 100 percent in Cuba and Panama, 84 percent in Nicaragua, 58 percent in Mexico, and 54 percent in Sri Lanka (FAO Fisheries and Aquaculture Department, 2014). While the monetary benefits of tilapia production in these environments are easily quantifiable, the potential trade-off in terms of losses of other ecosystem products or services that are likely to have occurred as a result of tilapia introduction and harvest is less well quantifiable. Previous evaluations of the ecosystem service trade-offs that could be generated by tilapia production have shown negative ecological impacts. For example, tilapia is thought to be the proximate cause of declines in many native fishes, including desert pupfish in the Southwest United States, Sinarapan from Lake Buhi in the Philippines, and various native cichlids. Other ecosystem services lost as a result of tilapia include the loss of aquatic plants and the habitats they provide for native species, as well as eutrophication-related biotic and abiotic alterations. Some tilapia fisheries, on the other hand, may help to protect local species by reducing fishing pressure on them. Because of the variety of reported ecological consequences and the lack of a quantitative evaluation that analyses the quality of inferences from various types of study designs, competing conceptions of tilapia's environmental implications have evolved. The lack of evidence for ecological consequences is likely to be substituted for evidence of the absence of ecological effects in management and policy decisions, in which case the monetized value of tilapia production may drive perceptions and decisions concerning tilapia introduction.

11.2 TILAPIA

Tilapia have already been utilized to suppress aquatic weeds, implying that these species may have a frequent impact on plant environments. Similarly, *O. mossambicus* is the sole cichlid included among the world's worst 100 invasive species, implying that *O. mossambicus* is more likely than other tilapia to have negative environmental impacts. The incidence and degree of ecological consequences from a species introduction are frequently influenced by the features of the community and ecosystem that receives it. Tilapia are native to Africa and a tiny section of the Middle East, but they may be found all over the world. It was projected that hybridization and competition between introduced and native tilapia species would be greatest in Africa because basic ecological theory predicts that these interactions will be strongest among similar species, despite recent evidence to the contrary. Mixed-model logistic regression was used to quantify the contribution of study design, species, and regions to the reported presence or absence of ecological consequences in order to answer the issues.

The discrepancy on tilapia impacts was addressed by separating the ecological implications of tilapia introduction, which are rarely marketed, from the more easily monetized ecosystem service provided by tilapia production and harvest. The geographical and socio-economic conditions of tilapia introduction, as well as ecological implications, all play a part in determining whether or not the fish is beneficial. While established ecological changes were substantially linked to unfavorable perceptions, the socioeconomic backdrop, particularly the value of tilapia production and hidden regional factors, also played a role. These findings support previous research indicating that invasion concepts are context specific and mutable, and that complex perceptions are common for introduced species that provide some ecosystem services while causing harm to others. Below are some broad considerations and potential assessments to help with decision-making and policy-making about tilapia introductions. This could be especially true for the United States and Sri Lanka, which are by far the two major contributors to tilapia research. These two countries account for 35% of global tilapia research papers, but only 2% of countries where tilapia populations have created feral populations. Furthermore, reports from these countries expressed widely divergent views on the ecosystem service outcomes of tilapia introduction.

In the United States, almost 80% of papers focused on negative consequences on ecosystem services, while in Sri Lanka, 75% of publications focused on positive results(Thushari & Senevirathna, 2020). Meanwhile, while Sri Lanka's habitat is ideal for tilapia, there are few, if any, native freshwater fish species that can withstand or be affected by introductions. Aside from livestock, fish is the most important source of animal protein. For fisheries, the wild fish resource is restricted. Fishermen, environmentalists, and policymakers are increasingly concerned about the depletion of natural supplies. Sustainable aquaculture has the potential to play a critical role in the transition to more environmentally and economically sustainable fish production. Land-based recirculating aquaculture and cage and offshore aquaculture provide a unique combination of wild stock conservation, socioeconomic benefits, and sustainability and scaling possibilities. Such technology has the potential to alter the way fish are produced, leading to large-scale fish culture similar to cattle, and hence aquaculture's future. Aquaculture, on the other hand, confronts a unique difficulty in that it relies on wild-caught fish as feed for breeding carnivorous fish, which may lead to further fish supply depletion. Aquaculture should add edible protein to the world rather than deplete it in the future. As a result, aquaculture species selection is crucial for ensuring a sustainable fish supply while also safeguarding wild fish supplies. Aquaculture's future will be defined by fish that do not require fish protein or require less fish protein in their diets.

Although tilapia farming has a lot of potential, there are a few obstacles to overcome. Compared to other fish such as salmon, farmed tilapias have low levels of important omega-3 fatty acids. Although some tilapia species such as Mozambique tilapia may grow in seawater, most tilapia species are not tolerant of high salinity. Tilapias' limited capacity to endure low temperatures limits the geographic breadth of tilapia production. To boost essential fatty acid content and tolerance to high salinity and cold, more research is needed. Selective breeding and enhanced diets using marine microalgae with high omega-3 concentration are likely to improve tilapia essential fatty acid levels. Salinity tolerance can be improved by crossing the fast-growing Nile tilapia with the salt-tolerant Mozambique tilapia. Marker-assisted selection and genomic selection will significantly expedite the genetic improvement of tilapias due to the rapid development of next-generation sequencing technology. The harmful effects of tilapia aquaculture on the environment and world biodiversity are another source of concern. To protect the ecosystem and aquatic biodiversity, proper monitoring and certification of tilapia farms are required. Land-based recirculating aquaculture, as well as cage and offshore culture, can lessen the negative effects of tilapia farming on the environment and world biodiversity, and are thus expected to rise in production and technological advancements. These methods make aquaculture possible practically anywhere, and will consequently be critical components of next-generation aquaculture.

Except for the fact that *O. mossambicus* had spread into numerous Asian countries by 1950, the origin of the first importation of tilapias into Asia is unknown. In most Asian countries, tilapias are now part of the fish assemblage. Tilapias have had an obvious impact on fish output in most nations, despite the fact that many governments do not officially declare their contribution to the FAO. It's rare, if not impossible, to find a country where tilapias haven't taken hold after their introduction. Despite their widespread introduction into Asian seas, there is no explicit evidence that tilapias have been too environmentally harmful, nor is there evidence that these species have a negative impact on biodiversity. In both the inland capture fishing and aquaculture sectors, tilapias continue to provide an essential and considerable contribution to fish output in Asia. This is an underestimation of productivity because several ASEAN countries have imported and are cultivating tilapias but do not report production to the FAO.

11.3 AQUACULTURE

Tilapias contribute to capturing fisheries throughout Asia and the Pacific's inland and lacustrine waterways or mostly reservoirs but have not been observed in rivers, with the exception of artisanal fishing in the Sepik River's floodplains in Papua New Guinea. Natural lakes are few in tropical Asia. Almost all lacustrine waterways in tropical Asia are reservoirs. From around 20°N latitude to about 15°S longitude, tilapia capture fisheries in lacustrine waterways have been observed in this region. Tilapias' contribution to overall landings in different water bodies, as well as their contribution to inland capture fisheries, varies greatly between water bodies and countries. Both in Africa and Asia and the Pacific, the inland tilapia capture fishery has doubled in the previous two decades and continues to grow. The Asia-Pacific region now accounts for roughly 20% of the global tilapia capture fisheries. For two main explanations, the yield from tilapia fisheries is likely under-estimated. Tilapia production might be combined with that of other cichlids or other freshwater fishes; and, some of the inland catch fishery produce could be used in aquaculture. Although it is clear that tilapia introductions are a key part of India's inland capture fisheries, species-level production statistics for tilapias are not included in the FAO's national data submissions. In Asia and the Pacific, however, the contribution of *O. niloticus* to the fisheries has recently decreased, whereas *O. mossambicus* has increased significantly. Despite the fact that the region's aggregate output has increased over time, there have been significant shifts in the contribution of individual countries, most notably Thailand. The Thai tilapia fishery began with the introduction of *O. niloticus*, peaked in the mid-to-late 1990s, and has since considerably dropped.

Because *Oreochromis mossambicus* was the first species to be introduced to Asia and the Pacific, it dominated inland fisheries in nations like Sri Lanka until recently. In Sri Lanka, the introduction of *O. mossambicus* encouraged the development of an inland fishery. Following that, *O. niloticus* was introduced into the region, and this species, as well as *O. mossambicus* x *O. niloticus* hybrids, progressively became the dominant species in most of Sri Lanka's reservoir fisheries. Inland fisheries in Sri Lanka fell in 1989, but are now rebounding. The primary cause of this declining trend is political, as a result of the Sri Lankan government's withdrawal of support and patronage for national aquaculture development in the late 1980s. For many years, however, tilapias have dominated Sri Lanka's inland fishery. Sri Lanka's inland catch fishery, which is almost exclusively dependent on introduced tilapias, is a one-of-a-kind fishery and one of the best-documented in the tropics. Even here, though, catch fishery production is not reported by species. While yields vary significantly amongst reservoirs, tilapias continue to account for the vast majority of landings in all of the various fisheries, accounting for roughly 25000 tonnes of a total production of almost 28000 tonnes. Variations in reservoir fish yields have been linked to land-use patterns in the reservoir catchment, with the more forest cover and grassland in the reservoir catchment in relation to reservoir area and/or capacity, the more productive the reservoir. Tilapia species accounted for more than 70% of the whole fishery in many cases. *O. mossambicus* had been introduced to several southern reservoirs by the end of the 1960s, and was contributing significantly to commercial catches in many of them. The paucity of reports makes estimating the contribution of tilapia to India's food security challenging. Even though the progressive domination of this species in the captures in several reservoirs was first observed in 1967 and reiterated in 1976, tilapia production from capture fisheries and aquaculture has not been recorded by FAO fisheries data systems. Tropical Indian reservoirs provide adequate habitat for *O. mossambicus*, which has formed self-sustaining populations in a number of south Indian reservoirs.

According to these findings, total fish output as well as tilapia production were significantly higher in smaller lakes. These two characteristics, however, did not appear to have any statistical link. In Tamil Nadu, for example, the average tilapia output in small lakes is estimated to be 53.68.2 kg/ha/yr. The introduction of tilapia to Indonesia has a long history. *O. mossambicus* is the dominant species in the Indonesian inland catch fishery, notably in natural lakes, as a result of its early introduction in the 1940s and subsequent spread to water bodies throughout the country. Tilapias were introduced to lakes in Indonesia, not all lakes had self-sustaining populations that contributed considerably to current fisheries. Nonetheless, tilapias account for more than 90% of total landings in some lakes and reservoirs. *O. niloticus* was the chosen species for stocking in newly developed reservoirs in Indonesia. As a result, this species is the mainstay of present reservoir fisheries. However, *O. mossambicus* or *O. niloticus* appear to be the dominant species in Indonesian reservoirs. Because *O. niloticus* was only introduced to Indonesia in 1969, it became the dominant species only in reservoirs erected in the 1970s, whereas *O. mossambicus* remains the dominating species in natural lakes.

11.4 PAPUA NEW GUINEA

Papua New Guinea (PNG) is located east of the Wallace and Weber lines, and hence represents a distinct zoogeographic zone from the rest of Asia. As a result, the freshwater fauna of the two regions differs significantly. The Australian region has created a freshwater ichthyo-fauna that originated from primarily marine families, making it less varied than mainland Asia. *Oreochromis mossambicus* supports a significant artisanal floodplain fishing in the Sepik River, accounting for over half of the landings, which are estimated to be between 3000 and 5000 tonnes per year. The Sepik's artisanal floodplain fishery was established and sustained(Glucksman, West, & Berra, 1976). The Government of Papua New Guinea, the United

Nations Development Program (UNDP), and the Food and Agriculture Organization (FAO) collaborated on a stock augmentation project in the Sepik and Ramu Rivers to provide greater food security to inland people. Carp, tilapia, and rainbow trout alien species had previously been imported for aquaculture development. Aquaculture, on the other hand, has struggled due to a lack of infrastructure and financial constraints. Tilapias have been classified as a noxious species in Australia. Tilapia introductions into natural and semi-natural waters in Australia were unintentional, and are assumed to have occurred as a result of aquarium imports. They were identified as self-propagating populations in tropical and sub-tropical Queensland and Western Australia. Feral tilapia populations are assumed to be made up of two species, *O. mossambicus* and *Tilapia mariae*, as well as a third morph, which is a possible interspecies hybrid between *O. mossambicus* and one or more species like *O. hornorum*, *O. niloticus*, and *O. aureus*.

11.5 INLAND CATCH TILAPIA FISHERIES

Inland catch tilapia fisheries (Figure 11.1) in the region are nearly entirely artisanal, relying on crafts like traditional dug-out boats or analogues. The most typically utilized equipment is gill nets. In the best of circumstances, individual craft landings rarely reach 100 kg every day. These fisheries are prone to daily changes in the catch. Landings, in the context of marketing, are much more localized and are not designed to handle vast quantities of produce. Because the numbers marketed on a daily basis in a given locality are generally in modest quantities, the marketing of inland tilapia catches has received little attention. Individual fishermen and primary vendors/wholesalers tend to be well-connected in these situations. The mechanism ensures that the food is sent quickly, even to the most remote locations. Tilapia is less expensive than the most popular marine but equivalent to the least expensive marine varieties. Tilapia prices are also competitive with those of other freshwater fish. Tilapias supply a relatively impoverished, rural community with an easily accessible and affordable animal protein source, and in some cases, they are the sole fish resource available to these populations.

In Sri Lanka, it is estimated that there are roughly 15000 full-time and part-time fishermen operating about 5,000 non-mechanized boats in perennial reservoirs(Pet & Piet, 1993). Because the island's interior fishery is based on tilapias, it's plausible to assume that this fishery provides direct work, and if the vendors who rely on the fisheries are factored in, the number of job opportunities provided increases dramatically. In Indonesia and the Philippines, where inland fisheries are primarily based on tilapias, the situation appears to be similar to Sri Lanka's in terms of creating jobs, increasing fish supplies at an affordable rate for the poor rural sectors of the community, and other socio-economic impacts.

Figure 11.1: Inland catch tilapia fisheries.

Source:https://www.fao.org/3/ae932e/ae932e0j.jpg

The tilapia is the most popular freshwater fish in the Philippines. As a result, the government assisted in improving its marketability and quality. The Philippine government's inland fisheries development policy is currently based on tilapias, which is also the chosen group for inland fishery enhancement.

Tilapias are critical to the survival of inland capture fisheries in Asia and the Pacific. Furthermore, tilapias provide an economical source of animal protein to poor, rural areas, in addition to contributing directly to fish supplies in Asia and the Pacific. Individual fisheries have emerged as a result of either planned seeding or invasion from other connecting waterways, and constant seeding is rare if ever, required to keep these fisheries alive. Despite the fact that only *O. mossambicus* and *O. niloticus* have been successful in contributing considerably to Asian inland capture fisheries, tilapias do not dominate fisheries in all Asian reservoirs and lakes. In contrast to Indonesia, the Philippines, and Sri Lanka, tilapias don't dominate the fisheries in Thailand. Indeed, the impact of tilapias on a fishery can vary greatly amongst water bodies and even within the same watershed. Despite the fact that tilapias have been present throughout Asia and the Pacific for more than five decades, there has been no documentation of sustainable populations established in the river systems. Since public awareness of the environmental risks of alien species has risen dramatically in the last decade, steps are being taken to better understand these impacts in order to mitigate them.

11.6 HYBRID VIGOR

A variety of variables have been credited with the region's tilapia fisheries' performance. The omnivorous feeding habits of introduced tilapias are the most notable of these, allowing for a shift in food preference based on availability and the ability to get necessary nutrients from food sources that appear to be nutritionally deficient. *Oreochromis niloticus* can also use blue-green algae as a nutritional source, which is abundant in tropical reservoirs during specific seasons, hence broadening the range of food sources available to these species. Tilapias' relatively high reproductive capacity i.e., ability to spawn throughout the year in most tropical freshwaters, low egg and fry mortality, availability of suitable nesting sites, short generation time, and other reproductive traits are all likely to contribute to their success in natural and semi-natural waters. The strong reproductive capacity could compensate for the relatively high fishing pressure and, in some cases, the very high bird predatory pressure on young, basically acting as a buffer. The success of tilapias was attributed to the availability of a vacant niche as a result of the lack of lacustrine species in the Asia-Pacific indigenous fish fauna(Xiao et al., 2022).

Hybrid vigor in the first generation may emerge from crossbreeding between species, resulting in an increase in production. However, this does not appear to be a major factor in tilapia colonization success. Hybridization between cohabiting tilapia species in natural or quasi-natural waterways is quite likely in Asia, as it is worldwide where tilapias have been introduced. Researchers found that hybrid fertility was much lower than that of the parent species. Some authors projected that in the long run, this would result in a decline in population reproductive capacity, which would be detrimental to the rather successful reservoir tilapia fishery in Sri Lanka reservoirs. To sustain strong fishing pressure and bird predation, tilapia populations in such reservoirs require high reproduction rates. Structure, and function by eliminating the number of species that are not adapted to the changing environment. The majority of countries that have introduced tilapias have signed the CBD and have agreed to safeguard them. Other causes, such as habitat degradation and overfishing, have been blamed for reductions in native biodiversity in many sites where tilapias have been introduced and cited as a problem. There is a significantly lower possibility that tilapias will pose a future threat to biodiversity if settings are well-maintained and ecosystems are intact. Thus, a decrease in native species could arise from direct interaction with an exotic species, increased fishing pressure, or changes in land use caused by the presence of a newly established species. When it comes to introductions, the immediate environment into which the organism is introduced is usually the center of attention. However, there is a chance that a significant environmental impact will occur in the future. The nearly complete extinction of the native cyprinid flock in Lake Lanao in the Philippines, where introduced tilapias are thought to have contributed to the reduction. The most particular negative consequences of tilapias have been linked to the near extinction of the little endemic goby in Lake Buhi, Philippines.

It's also been reported that the introduction of *O. mossambicus* to Lake Toba, Indonesia, caused the decline of the indigenous cyprinid species Lissochilus, which is regarded culturally important. The latter is thought to account for 86 percent of the lake's fish landings. There is no supporting evidence for this observation. A similar situation has been observed at Kaptai Lake, a huge reservoir in Bangladesh, where an increase in *O. niloticus* landings has resulted in the loss of the local carp fishery. However, *O. niloticus* landings account for less than 1% of the reservoir's overall fish yield. There have been a few sporadic cases when tilapia invasions have been linked to specific faunal alterations. The introduced tilapias, while present at the location, are most likely not the primary source of the modifications in these cases. Introduced tilapias are a potential threat to Sri Lanka's native freshwater fish. Tilapias are not found in Sri Lanka's streams, or other natural waterways. However, it was proposed that as lowland streams become more silted as a result of increased deforestation, tilapias could eventually infiltrate the central hills. The main breeding season of *O. mossambicus* coincides with that of most indigenous species, implying that the latter has been displaced. Tilapias and indigenous species coexist without conflict for food resources, and tilapias do not graze on the young of any indigenous species. It was also argued that excessive fishing pressure for *O. mossambicus* in reservoirs contributed to the decline of freshwater turtles, *Lissemys punctata* and *Melanochelys trijuga*.

The inland fish fauna of Australia has a high level of endemism. However, comprehensive studies have shown that *O. mossambicus* has had no ecological influence in Australia at this time. When they exist together, for example, it was found that it does not compete for food resources with indigenous eel-tailed catfish, *Tandanus tandanus*, and the omnivorous teraponid, *Leiopotherapon unicolor*. The researchers anticipated that a few more decades would be required before any strong conclusions could be reached about its effects. Considering the evidence presented, it is reasonable to infer that purposeful and unintentional tilapia imports to Asia and the Pacific have resulted in overall good outcomes. Tilapias are a valuable fishery resource in most countries in the region, where they provide jobs and provide a low-cost, easily accessible animal protein source for the poorest members of society. There is no clear evidence or objective synthesis of currently available data to demonstrate that tilapias have caused detrimental ecological repercussions in the region, particularly the loss of biodiversity. It is vital to maintain constant vigilance and impartial assessments of their impact on the ecosystem and biodiversity. Monitoring and objective continual assessments are also required. The current success of tilapia introductions should not prevent countries from proceeding with caution and responsibility in the future, adopting international rules of practice such as the ICES/EIFAC Code and guidelines. In contrast to the good effects of tilapia introductions in Asia's mainland states and Papua New Guinea, tilapia introductions to South Pacific Island nations are still disputed. Only *O. mossambicus* that is introduced to Cook Islands, Samoas, Wallis Island, Fiji, New Caledonia, Solomon Islands, Tonga, Line Islands, and Caroline Islands at Yap and *T. zillii* are known to have established themselves in natural and/

or quasi-natural freshwater habitats. In Papua New Guinea, Tonga, Caroline Islands, Nauru, and Kiribati, tilapia species were established in brackish water habitats. *O. mossambicus* was accidentally introduced to Fanning Atoll in the Linne Islands in 1959, resulting in its spread to a number of lagoons on the atoll. There have been no known negative effects on the lagoon's marine species. The atoll's population does not eat tilapias, but they are regarded to be a vital food source for carnivorous marine species that migrate in and out of the lagoons with the tides.

The world's tilapia cultivation procedures are extremely diverse, arguably the most different of all aquaculture species. It is a group of fish that can be grown at a variety of intensities, making it appealing to people from diverse socioeconomic backgrounds and allowing culture techniques to be tailored to their financial capacities. *Oreochromis niloticus* is often cultured in backyards and/or home garden ponds to supplement income and give a fresh source of animal proteins for the family. In these cases, the cultured stock is frequently fed kitchen garbage and supplemented with readily available, often low-cost agricultural by-products like rice bran. However, the latter's direct nutritional benefit to the stock is unknown, and it is likely to be modest; the inputs are more of a fertilizer. *Oreochromis niloticus* is cultivated in sewage-fed ponds and primary and secondary treated waste effluents. Despite the fact that the method has been in place since the 1930s, there have been no reports of negative health impacts from eating fish raised on sewage-fed farms. Commercial feed makers in Asia's developing countries frequently see tilapias as a good fish species for marketing their feeds through community-based subsidy programs. The robust properties of tilapia, such as its high growth rate and relative disease resistance, appeal to rural farmers who are new to aquaculture and would be prepared to accept such schemes. With the current trend of increased rivalry among industrial producers of aquaculture commodities, especially feeds, such feed subsidy programs are anticipated to grow popular among rural communities, contributing to overall output gains.

Tilapia cage culture can offer alternative livelihood to those displaced by inland dam construction and reservoir impoundment. Despite growing community opposition to dam construction, particularly across important rivers, it continues uninterrupted in most developing countries for reasons other than fisheries. Many people believe that tilapia cage culture could provide an alternative source of income for displaced people, and that such programs could help boost overall tilapia production in the coming years. Twenty-five years ago, forecasts were made that sewage treatment plants would become big food production centers and that excreta utilization in aquaculture would become an increasingly important source of waste disposal and food production. Night soil is utilized for agricultural output, though this practice has been contentious due to food safety concerns and is steadily disappearing. There will be greater demand for waste management, and the recycling of biological wastes for food production, both directly and indirectly, will become increasingly important. Tilapias have shown to be excellent candidates for this task. Tilapias are currently cultured in sewage ponds and wastewater treatment plants, as previously stated. With growing pressures on communities to recycle nutritional resources, such operations have a real chance of expanding and adding to overall farmed tilapia production in the tropics. Tilapia culture will continue to be an appealing proposition for both poor communities and those who are embarking on aquaculture, either to supplement household income or as an alternative livelihood. Tilapia has some benefits over other species produced in intensive systems because of its omnivorous feeding habit, which requires less protein. Despite significant progress in laboratory research on the partial substitution of fish meal in tilapia diets, no farmed fish species has ever proved that effective diets can be made without fish meal. Fish meal must continue to be included in feeds. Future tilapia culture expansion will be fueled mostly by export markets to the United States of America.

America, Japan, and a few European countries, as well as the Middle East, are all destinations. Chemical and veterinary medicine residues must meet high-quality standards in these countries. As a result, in order to remain competitive, developing-country manufacturers must meet these requirements and strive to provide high-quality, safe-to-eat products that meet consumer demand and expectations. Tilapia culture methods will require a general improvement in facilities and attitudes in dealing with a commodity that is not deemed high value, profit margins are rather narrow, and market demand could be a limitation. In nations where tilapias have been introduced, there are growing problems about potential damage to biodiversity. It was also demonstrated that the majority, if not all, of the apparent effects of tilapias on biodiversity, are attributable to factors other than tilapias, and explicit proof in this regard is lacking. Tilapia can harm natural biodiversity in a variety of ways. They upset the natural balance and have a negative impact on the local ecosystem by contending for food and shelter, and tilapia extensive farming causes water to become eutrophicated. They gradually reduce the diversity and density of indigenous plants, which are used by local fish species for spawning and protection. Tilapia feeds mostly on the eggs of other species, causing local species to become extinct. They not only harm local species, but they also spread disease throughout aquatic ecosystems. In addition to commercial fisheries, the invasion of Nile tilapia had a significant impact on recreational fishing. The phytoplankton and zooplankton composition in ponds and lakes determines primary productivity. The invasion of tilapia, which graze on them, reduces production. Nile tilapia can diminish phytoplankton and zooplankton abundance, composition, and biomass in tropical lakes and reservoirs, however phytoplankton biomass and size-structure impact the extent of the effects.

REFERENCES

1. Abdel-Khalek, A. A., Zayed, H. S., Elsayad, S. M., & Zaghloul, K. H. (2020). Assessment of metal pollution impacts on Tilapia zillii and Mugil cephalus inhabiting Qaroun and Wadi El-Rayan lakes, Egypt, using integrated biomarkers. Environmental Science and Pollution Research, 27(21), 26773–26785. https://doi.org/10.1007/s11356-020-09095-3
2. Abdel-Latif, H. M. R., Dawood, M. A. O., Menanteau-Ledouble, S., & El-Matbouli, M. (2020). The nature and consequences of co-infections in tilapia: A review. Journal of Fish Diseases, 43(6), 651–664. https://doi.org/10.1111/jfd.13164
3. Acosta, B. O., & Gupta, M. V. (2010). The Genetic Improvement of Farmed Tilapias Project: Impact and Lessons Learned. In S. S. De Silva & F. B. Davy (Eds.), Success Stories in Asian Aquaculture (pp. 149 171). Dordrecht: Springer Netherlands. https://doi.org/10.1007/978-90-481-3087-0_8
4. Ahmed, B. N., & Waibel, H. (2019). The role of homestead fish ponds for household nutrition security in Bangladesh. Food Security, 11(4), 835–854. https://doi.org/10.1007/s12571-019-00947-6
5. Ahmed, N., Ahammed, F., & Lecouffe, C. (2007). Socioeconomic Aspects of Small-Scale Freshwater Prawn Marketing Systems in Mymensingh, Bangladesh. Aquaculture Economics & Management, 11(4), 335–353. https://doi.org/10.1080/13657300701530324

6. Ahmed, N., Young, J. A., Dey, M. M., & Muir, J. F. (2012). From production to consumption: A case study of tilapia marketing systems in Bangladesh. Aquaculture International, 20(1), 51–70. https://doi.org/10.1007/s10499-011-9441-0
7. Aich, N., Paul, A., Choudhury, T. G., & Saha, H. (2022). Tilapia Lake Virus (TiLV) disease: Current status of understanding. Aquaculture and Fisheries, 7(1), 7–17. https://doi.org/10.1016/j.aaf.2021.04.007
8. Alam, A., Chadha, N. K., Joshi, K. D., Chakraborty, S. K., Sawant, P. B., Kumar, T., … Sharma, A. P. (2015). Food and Feeding Ecology of the Non-native Nile Tilapia Oreochromis niloticus (Linnaeus, 1758) in the River Yamuna, India. Proceedings of the National Academy of Sciences, India Section B: Biological Sciences, 85(1), 167–174. https://doi.org/10.1007/s40011-014-0338-3
9. Alam, Md. A., Guttormsen, A. G., & Roll, K. H. (2019). Production Risk and Technical Efficiency of Tilapia Aquaculture in Bangladesh. Marine Resource Economics, 34(2), 123–141. https://doi.org/10.1086/704129
10. Ali, H., & Haque, M. M. (2011). Impacts of Pangasius aquaculture on land use patterns in Mymensingh district of Bangladesh. Journal of the Bangladesh Agricultural University, 9(1), 169–170. Retrieved from https://www.banglajol.info/index.php/JBAU/article/view/8759
11. Ali, Hazrat, Rico, A., Murshed-e-Jahan, K., & Belton, B. (2016). An assessment of chemical and biological product use in aquaculture in Bangladesh. Aquaculture, 454, 199–209. https://doi.org/10.1016/j.aquaculture.2015.12.025
12. Ali, M. S., Griffiths, D., Turner, W., & Farms, N. S. (n.d.). Practical training – Tilapia breeding and all-male fry production. 29.
13. Amarasinghe, U. S. (1987). Status of the fishery of Pimbmrettewa wewa, a man-made lake in Sri Lanka. Aquaculture Research, 18(4), 375–385. https://doi.org/10.1111/j.1365-2109.1987.tb00326.x
14. Amarasinghe, U. S. (1990). Minor cyprinid resources in a man-made lake in Sri Lanka: A potential supplementary source of income for fishermen. Fisheries Research, 9(1), 81–89. https://doi.org/10.1016/0165-7836(90)90043-U
15. Anane-Taabeah, G., Quagrainie, K., & Amisah, S. (2016). Assessment of farmed tilapia value chain in Ghana. Aquaculture International, 24(4), 903–919. https://doi.org/10.1007/s10499-015-9960-1

16. Apud, F., Primavera, J. H., & Torres, P. L. (1983). Farming of prawns and shrimps. (5), 78.

17. Arthur, R. I., Lorenzen, K., Homekingkeo, P., Sidavong, K., Sengvilaikham, B., & Garaway, C. J. (2010). Assessing impacts of introduced aquaculture species on native fish communities: Nile tilapia and major carps in SE Asian freshwaters. Aquaculture, 299(1), 81–88. https://doi.org/10.1016/j.aquaculture.2009.11.022

18. Aswathy, N., Imelda, J., Ignatius, B., & Joseph, S. (2020). Economic viability of cage fish farming in India (No. 134). Kochi: Central Marine Fisheries Research Institute. Retrieved from http://eprints.cmfri.org.in/14274/

19. Athukoorala, A. A. S. H., Bhujel, R. C., Krakstad, J.-O., & Matsuishi, T. F. (2021). Regional variation in fish species on the continental shelf of Sri Lanka. Regional Studies in Marine Science, 44, 101755. https://doi.org/10.1016/j.rsma.2021.101755

20. Bacharach, E., Mishra, N., Briese, T., Zody, M. C., Kembou Tsofack, J. E., Zamostiano, R., … Lipkin, W. I. (n.d.). Characterization of a Novel Orthomyxo-like Virus Causing Mass Die-Offs of Tilapia. MBio, 7(2), e00431-16. https://doi.org/10.1128/mBio.00431-16

21. Bainy, E. M., Bertan, L. C., Corazza, M. L., & Lenzi, M. K. (2015). Effect of grilling and baking on physicochemical and textural properties of tilapia (Oreochromis niloticus) fish burger. Journal of Food Science and Technology, 52(8), 5111–5119. https://doi.org/10.1007/s13197-014-1604-3

22. Baki, B., & Yucel, Ş. (2017). Feed Cost/production Income Analysis Of Seabass (Dicentrarchus labrax) Aquaculture.

23. Bandara, J. M. R. S., Senevirathna, D. M. A. N., Dasanayake, D. M. R. S. B., Herath, V., Bandara, J. M. R. P., Abeysekara, T., & Rajapaksha, K. H. (2008). Chronic renal failure among farm families in cascade irrigation systems in Sri Lanka associated with elevated dietary cadmium levels in rice and freshwater fish (Tilapia). Environmental Geochemistry and Health, 30(5), 465–478. https://doi.org/10.1007/s10653-007-9129-6

24. Baroiller, J. F., D'Cotta, H., Bezault, E., Wessels, S., & Hoerstgen-Schwark, G. (2009). Tilapia sex determination: Where temperature and genetics meet. Comparative Biochemistry and Physiology Part A: Molecular & Integrative Physiology, 153(1), 30–38. https://doi.org/10.1016/j.cbpa.2008.11.018

25. Basri, L., Nor, R. M., Salleh, A., Md. Yasin, I. S., Saad, M. Z., Abd. Rahaman, N. Y., … Amal, M. N. A. (2020). Co-Infections of Tilapia Lake Virus, Aeromonas hydrophila and Streptococcus agalactiae in Farmed Red Hybrid Tilapia. Animals, 10(11), 2141. https://doi.org/10.3390/ani10112141
26. Behera, B. K., Pradhan, P. K., Swaminathan, T. R., Sood, N., Paria, P., Das, A., … Jena, J. K. (2018). Emergence of Tilapia Lake Virus associated with mortalities of farmed Nile Tilapia Oreochromis niloticus (Linnaeus 1758) in India. Aquaculture, 484, 168–174. https://doi.org/10.1016/j.aquaculture.2017.11.025
27. Belton, B., Karim, M., Thilsted, S., Murshed-E-Jahan, K., Collis, W., & Phillips, M. (2011). Review of aquaculture and fish consumption in Bangladesh. Http://Aquaticcommons.Org/Id/Eprint/7516. Retrieved from https://aquadocs.org/handle/1834/24444
28. Belton, Ben, Little, D., & Grady, K. (2009). Is Responsible Aquaculture Sustainable Aquaculture? WWF and the Eco-Certification of Tilapia. Society & Natural Resources, 22(9), 840–855. https://doi.org/10.1080/08941920802506257
29. Belton, Ben, & Thilsted, S. H. (2014). Fisheries in transition: Food and nutrition security implications for the global South. Global Food Security, 3(1), 59–66. https://doi.org/10.1016/j.gfs.2013.10.001
30. Bentsen, H. B., Gjerde, B., Eknath, A. E., de Vera, M. S. P., Velasco, R. R., Danting, J. C., … Ponzoni, R. W. (2017). Genetic improvement of farmed tilapias: Response to five generations of selection for increased body weight at harvest in Oreochromis niloticus and the further impact of the project. Aquaculture, 468, 206–217. https://doi.org/10.1016/j.aquaculture.2016.10.018
31. Berg, H., Mulokozi, D., & Udikas, L. (2021). A GIS Assessment of the Suitability of Tilapia and Clarias Pond Farming in Tanzania. ISPRS International Journal of Geo-Information, 10(5), 354. https://doi.org/10.3390/ijgi10050354
32. Bhujel, R. (2008). The Role Of The Asian Institute Of Technology In The Promotion Of Tilapia For Aquaculture.
33. Bhujel, R. C. (2014). A Manual for Tilapia Business Management. CABI.
34. Bigarré, L., Cabon, J., Baud, M., Heimann, M., Body, A., Lieffrig, F., & Castric, J. (2009). Outbreak of betanodavirus infection in tilapia,

Oreochromis niloticus (L.), in fresh water. Journal of Fish Diseases, 32(8), 667–673. https://doi.org/10.1111/j.1365-2761.2009.01037.x

35. Bogard, J. R., Farook, S., Marks, G. C., Waid, J., Belton, B., Ali, M., … Thilsted, S. H. (2017). Higher fish but lower micronutrient intakes: Temporal changes in fish consumption from capture fisheries and aquaculture in Bangladesh. PLOS ONE, 12(4), e0175098. https://doi.org/10.1371/journal.pone.0175098
36. Booncharoen, C., & Anal, A. K. (2021). Attitudes, Perceptions, and On-Farm Self-Reported Practices of Shrimp Farmers' towards Adoption of Good Aquaculture Practices (GAP) in Thailand. Sustainability, 13(9), 5194. https://doi.org/10.3390/su13095194
37. Boyd, C. E. (n.d.). Farm-Level Issues in Aquaculture Certification: Tilapia. 29.
38. Cai, J., Zhou, X., Yan, X., Lucente, D., & Lagana, C. (2019). Top 10 species groups in global aquaculture 2017. 12.
39. Canevari-Luzardo, L. M., Berkhout, F., & Pelling, M. (2020). A relational view of climate adaptation in the private sector: How do value chain interactions shape business perceptions of climate risk and adaptive behaviours? Business Strategy and the Environment, 29(2), 432–444. https://doi.org/10.1002/bse.2375
40. Carp Cultivation and the Highlands Aquaculture Development Centre, Aiyura. (n.d.). Retrieved April 27, 2022, from https://www.fao.org/3/r9972e/r9972e10.htm
41. Chakraborty, S. B., Mazumdar, D., & Banerjee, S. (2010). Determination of ideal stocking density for cage culture of monosex Nile tilapia (Oreochromis niloticus) in India. Proceedings of the Zoological Society, 63(1), 53–59. https://doi.org/10.1007/s12595-010-0007-3
42. Chalermwat, K., Szuster, B. W., & Flaherty, M. (2003). Shellfish aquaculture in Thailand. Aquaculture Economics & Management, 7(3–4), 249–261. https://doi.org/10.1080/13657300309380343
43. Chatvijitkul, S., Boyd, C. E., Davis, D. A., & McNevin, A. A. (2017). Pollution potential indicators for feed-based fish and shrimp culture. Aquaculture, 477, 43–49. https://doi.org/10.1016/j.aquaculture.2017.04.034
44. Chen, C.-Y., Wooster, G. A., & Bowser, P. R. (2004). Comparative blood chemistry and histopathology of tilapia infected with Vibrio vulnificus or Streptococcus iniae or exposed to carbon tetrachloride,

gentamicin, or copper sulfate. Aquaculture, 239(1), 421–443. https://doi.org/10.1016/j.aquaculture.2004.05.033

45. Chenyambuga, S. W., Madalla, N., & Mnembuka, B. (n.d.). Management and Value Chain of Nile Tilapia Cultured in Ponds of Small-scale Farmers in Morogoro Region, Tanzania.
46. Chiasson, M., & Benfey, T. (2007). Gonadal differentiation and hormonal sex reversal in Arctic charr (Salvelinus alpinus). Journal of Experimental Zoology. Part A, Ecological Genetics and Physiology, 307, 527–534. https://doi.org/10.1002/jez.407
47. Chitmanat, C., Lebel, P., Whangchai, N., Promya, J., & Lebel, L. (2016). Tilapia diseases and management in river-based cage aquaculture in northern Thailand. Journal of Applied Aquaculture, 28(1), 9–16. https://doi.org/10.1080/10454438.2015.1104950
48. Chumnongsittathum, B. (2008). Community-based aquaculture for poverty alleviation and sustainable livelihoods. In Handbook on Community-based Aquaculture for Remote Rural Areas of Southeast Asia (pp. 6–13). Secretariat, Southeast Asian Fisheries Development Center. Retrieved from http://hdl.handle.net/20.500.12066/4644
49. Conte, M. A., Gammerdinger, W. J., Bartie, K. L., Penman, D. J., & Kocher, T. D. (2017). A high quality assembly of the Nile Tilapia (Oreochromis niloticus) genome reveals the structure of two sex determination regions. BMC Genomics, 18(1), 341. https://doi.org/10.1186/s12864-017-3723-5
50. Costa-Pierce, B. A. (2003). Rapid evolution of an established feral tilapia (Oreochromis spp.): The need to incorporate invasion science into regulatory structures. Biological Invasions, 5(1), 71–84. https://doi.org/10.1023/A:1024094606326
51. Dan, N. C., & Little, D. C. (2000). Overwintering performance of Nile tilapia Oreochromis niloticus (L.) broodfish and seed at ambient temperatures in northern Vietnam. Aquaculture Research, 31(6), 485–493. https://doi.org/10.1046/j.1365-2109.2000.00466.x
52. Dasgupta, S., Mustafa, G., Paul, T., & Wheeler, D. (2021). The socioeconomics of fish consumption and child health: An observational cohort study from Bangladesh. World Development, 137, 105201. https://doi.org/10.1016/j.worlddev.2020.105201
53. Debnath, P. P., Dinh-Hung, N., Taengphu, S., Nguyen, V. V., Delamare-Deboutteville, J., Senapin, S., … Rodkhum, C. (2022). Tilapia Lake

Virus was not detected in non-tilapine species within tilapia polyculture systems of Bangladesh. Journal of Fish Diseases, 45(1), 77–87. https://doi.org/10.1111/jfd.13537

54. Deines, A. M., Wittmann, M. E., Deines, J. M., & Lodge, D. M. (2016). Tradeoffs among Ecosystem Services Associated with Global Tilapia Introductions. Reviews in Fisheries Science & Aquaculture, 24(2), 178–191. https://doi.org/10.1080/23308249.2015.1115466

55. del-Pozo, J., Mishra, N., Kabuusu, R., Cheetham, S., Eldar, A., Bacharach, E., … Ferguson, H. W. (2017). Syncytial Hepatitis of Tilapia (Oreochromis niloticus L.) is Associated With Orthomyxovirus-Like Virions in Hepatocytes. Veterinary Pathology, 54(1), 164–170. https://doi.org/10.1177/0300985816658100

56. Dey, M. M., & Eknath, A. E. (1997). Current trends in the Asian tilapia industry and the significance of genetically improved tilapia breeds. In Sustainable aquaculture. INFOFISH. Retrieved from https://digitalarchive.worldfishcenter.org/handle/20.500.12348/2638

57. Dey, Madan M., Alam, M. F., & Bose, M. L. (2010). Demand for aquaculture development: Perspectives from Bangladesh for improved planning. Reviews in Aquaculture, 2(1), 16–32. https://doi.org/10.1111/j.1753-5131.2010.01020.x

58. Dey, Madan M., Bimbao, G. B., Yong, L., Regaspi, P., Kohinoor, A. H. M., Pongthana, N., & Paraguas, F. J. (2000). Current status of production and consumption of tilapia in selected Asian countries. Aquaculture Economics & Management, 4(1–2), 13–31. https://doi.org/10.1080/13657300009380258

59. Dey, Madan M., & Gupta, M. V. (2000). Socioeconomics of disseminating genetically improved Nile tilapia in Asia: An introduction. Aquaculture Economics & Management, 4(1–2), 5–11. https://doi.org/10.1080/13657300009380257

60. Dinh-Hung, N., Sangpo, P., Kruangkum, T., Kayansamruaj, P., Rung-ruangkijkrai, T., Senapin, S., … Dong, H. T. (2021). Dissecting the localization of Tilapia tilapinevirus in the brain of the experimentally infected Nile tilapia, Oreochromis niloticus (L.). Journal of Fish Diseases, 44(8), 1053–1064. https://doi.org/10.1111/jfd.13367

61. Dong, H. T., LaFrentz, B., Pirarat, N., & Rodkhum, C. (2015). Phenotypic characterization and genetic diversity of Flavobacterium columnare isolated from red tilapia, Oreochromis sp., in Thailand.

Journal of Fish Diseases, 38(10), 901–913. https://doi.org/10.1111/jfd.12304

62. Dong, H. T., Siriroob, S., Meemetta, W., Santimanawong, W., Gangnonngiw, W., Pirarat, N., … Senapin, S. (2017). Emergence of tilapia lake virus in Thailand and an alternative semi-nested RT-PCR for detection. Aquaculture, 476, 111–118. https://doi.org/10.1016/j.aquaculture.2017.04.019
63. Eknath, A. E., & Hulata, G. (2009). Use and exchange of genetic resources of Nile tilapia (Oreochromis niloticus). Reviews in Aquaculture, 1(3–4), 197–213. https://doi.org/10.1111/j.1753-5131.2009.01017.x
64. Fan, L., Chen, J., Meng, S., Song, C., Qiu, L., Hu, G., & Xu, P. (2017). Characterization of microbial communities in intensive GIFT tilapia (Oreochromis niloticus) pond systems during the peak period of breeding. Aquaculture Research, 48(2), 459–472. https://doi.org/10.1111/are.12894
65. Fisheries and Aquaculture—Cultured Aquatic Species—Oreochromis niloticus. (n.d.). Retrieved April 27, 2022, from https://www.fao.org/fishery/en/culturedspecies/oreochromis_niloticus?lang=en
66. Glucksman, J., West, G., & Berra, T. M. (1976). The introduced fishes of Papua New Guinea with special reference to Tilapia mossambica. Biological Conservation, 9(1), 37–44. https://doi.org/10.1016/0006-3207(76)90072-0
67. Goes, E. S. R., Lara, J. A. F., Gasparino, E., Del Vesco, A. P., Goes, M. D., Alexandre Filho, L., & Ribeiro, R. P. (2015). Pre-Slaughter Stress Affects Ryanodine Receptor Protein Gene Expression and the Water-Holding Capacity in Fillets of the Nile Tilapia. PLoS ONE, 10(6), e0129145. https://doi.org/10.1371/journal.pone.0129145
68. Groenewald, N. J. (n.d.). Comparisons of growth performance of Nile tilapia (Oreochromis niloticus) fingerlings fed different inclusion levels of black soldier fly (Hermetia illucens) pre-pupae meal diets and its effect on the physical characteristics of the feed. 91.
69. Gupta, D., Dwivedi, A. K., & Tripathi, M. (2018). Taxonomic validation of five fish species of subfamily Barbinae from the Ganga river system of northern India using traditional and truss analyses. PLOS ONE, 13(10), e0206031. https://doi.org/10.1371/journal.pone.0206031

70. Guz, L., Puk, K., Walczak, N., Oniszczuk, T., & Oniszczuk, A. (2014). Effect of dietary supplementation with Echinacea purpurea on vaccine efficacy against infection with Flavobacterium columnare in zebrafish (Danio rerio). Polish Journal of Veterinary Sciences, 17(4). Retrieved from https://bibliotekanauki.pl/articles/32291

71. Hamilton, M. G., Lind, C. E., Barman, B. K., Velasco, R. R., Danting, Ma. J. C., & Benzie, J. A. H. (2020). Distinguishing Between Nile Tilapia Strains Using a Low-Density Single-Nucleotide Polymorphism Panel. Frontiers in Genetics, 11. Retrieved from https://www.frontiersin.org/article/10.3389/fgene.2020.594722

72. Haridas, H., Verma, A. K., Rathore, G., Prakash, C., Sawant, P. B., & Babitha Rani, A. M. (2017). Enhanced growth and immuno-physiological response of Genetically Improved Farmed Tilapia in indoor biofloc units at different stocking densities. Aquaculture Research, 48(8), 4346–4355. https://doi.org/10.1111/are.13256

73. Harkes, I. H. T., Drengstig, A., Kumara, M. P., Jayasinghe, J. M. P. K., & Huxham, M. (2015). Shrimp aquaculture as a vehicle for Climate Compatible Development in Sri Lanka. The case of Puttalam Lagoon. Marine Policy, 61, 273–283. https://doi.org/10.1016/j.marpol.2015.08.003

74. Harohau, D., Sulu, R. J., Phillips, M. J., Sukulu, M., Pickering, T., & Schwarz, A. M. (2016). Improving household tilapia (Oreochromis mossambicus) aquaculture through participatory action research. Aquaculture, 465, 272–286. https://doi.org/10.1016/j.aquaculture.2016.09.024

75. Hirimuthugoda, N. Y., Liyanage, U., & Amaraweera, K. W. R. R. (2014). Polyculture of Giant freshwater prawn (Macrobrachium rosenbergii) and Nile tilapia (Oreochromis niloticus) in earthen ponds at different stocking densities [Conference poster]. ISAE 2014. Proceedings of the International Symposium on Agriculture and Environment 2014, Ruhuna, Sri Lanka, 27 November 2014, 313–318. Retrieved from https://www.cabdirect.org/cabdirect/abstract/20153384261

76. Hua, K., Cobcroft, J. M., Cole, A., Condon, K., Jerry, D. R., Mangott, A., … Strugnell, J. M. (2019). The Future of Aquatic Protein: Implications for Protein Sources in Aquaculture Diets. One Earth, 1(3), 316–329. https://doi.org/10.1016/j.oneear.2019.10.018

77. Hussain, M. G. (2004). Farming of Tilapia: Breeding Plans, Mass Seed Production and Aquaculture Techniques.

78. Ibrahim, N., & Naggar, G. E. (2010). Water Quality, Fish Production and Economics of Nile Tilapia, Oreochromis niloticus, and African Catfish, Clarias gariepinus, Monoculture and Polycultures. Journal of the World Aquaculture Society, 41(4), 574–582. https://doi.org/10.1111/j.1749-7345.2010.00397.x

79. Jansen, M. D., Dong, H. T., & Mohan, C. V. (2019). Tilapia lake virus: A threat to the global tilapia industry? Reviews in Aquaculture, 11(3), 725–739. https://doi.org/10.1111/raq.12254

80. Jantrakajorn, S., Maisak, H., & Wongtavatchai, J. (2014). Comprehensive Investigation of Streptococcosis Outbreaks in Cultured Nile Tilapia, Oreochromis niloticus, and Red Tilapia, Oreochromis sp., of Thailand. Journal of the World Aquaculture Society, 45(4), 392–402. https://doi.org/10.1111/jwas.12131

81. Jayasankar, P. (2018). Present status of freshwater aquaculture in India—A review. Indian Journal of Fisheries, 65(4), 157–165. Retrieved from http://epubs.icar.org.in/ejournal/index.php/IJF/issue/view/2488

82. Jiménez-Badillo, L. (2006). Age-growth models for tilapia Oreochromis aureus (Perciformes, Cichlidae) of the Infiernillo reservoir, Mexico and reproductive behaviour. Revista de Biología Tropical, 54(2), 577–588. Retrieved from http://www.scielo.sa.cr/scielo.php?script=sci_abstract&pid=S0034-77442006000200031&lng=en&nrm=iso&tlng=en

83. Joshi, R., Árnyasi, M., Lien, S., Gjøen, H. M., Alvarez, A. T., & Kent, M. (2018). Development and Validation of 58K SNP-Array and High-Density Linkage Map in Nile Tilapia (O. niloticus). Frontiers in Genetics, 9. Retrieved from https://www.frontiersin.org/article/10.3389/fgene.2018.00472

84. Kabir, M. H., & Eva, I. J. (2014). Environmental impacts of shrimp aquaculture: The case of Chandipur village at Debhata upazila of Satkhira district, Bangladesh. Journal of the Asiatic Society of Bangladesh, Science, 40(1), 107–119. https://doi.org/10.3329/jasbs.v40i1.31738

85. Kamaruzzaman, N., Nguyen, N. H., Hamzah, A., & Ponzoni, R. W. (2009). Growth performance of mixed sex, hormonally sex reversed and progeny of YY male tilapia of the GIFT strain, Oreochromis niloticus.

Aquaculture Research, 40(6), 720–728. https://doi.org/10.1111/j.1365-2109.2008.02152.x

86. Karim, M., Ahmed, M., Talukder, R. K., Taslim, M. A., & Rahman, H. Z. (2006). Policy working paper: Dynamic agribusiness-focused aquaculture for poverty reduction and economic growth in Bangladesh [Working Paper]. WorldFish Center. Retrieved from WorldFish Center website: https://digitalarchive.worldfishcenter.org/handle/20.500.12348/1867

87. Kassam, L. (n.d.). Aquaculture and food security, poverty alleviation and nutrition in Ghana: Case study prepared for the Aquaculture for Food Security, Poverty Alleviation and Nutrition project. 47.

88. Kibenge, F. S. (2019). Emerging viruses in aquaculture. Current Opinion in Virology, 34, 97–103. https://doi.org/10.1016/j.coviro.2018.12.008

89. Kuo, C. M., & Neal, R. A. (1982). ICLARM's tilapia research. ICLARM Newsletter. Retrieved from https://digitalarchive.worldfishcenter.org/handle/20.500.12348/3587

90. Kuwaye, T. T., Okimoto, D. K., Shimoda, S. K., Howerton, R. D., Lin, H.-R., Pang, P. K. T., & Grau, E. G. (1993). Effect of 17α-methyltestosterone on the growth of the euryhaline tilapia, Oreochromis mossambicus, in fresh water and in sea water. Aquaculture, 113(1), 137–152. https://doi.org/10.1016/0044-8486(93)90347-2

91. Lakpawan, W. M. P., Liyanage, N. P. P., Tennakoon, T. M. P. S. I., Herath, H. M. T. N. B., & Kulathilaka, C. J. P. (2019). A Comparative Study on Ancient and Contemporary Fisheries Management Systems in Inland Reservoirs of Anuradhapura District in Sri Lanka: A Review. Uva Wellassa University of Sri Lanka. Retrieved from http://www.erepo.lib.uwu.ac.lk/handle/123456789/62

92. Lalramchhani, C., Balasubramanian, C. P., Panigrahi, A., Ghoshal, T. K., Das, S., Shyne Anand, P. S., & Vijayan, K. K. (2019). Polyculture of Indian White Shrimp (Penaeus indicus) with Milkfish (Chanos chanos) and its Effect on Growth Performances, Water Quality and Microbial Load in Brackishwater Pond. Journal of Coastal Research, 86(SI), 43–48. https://doi.org/10.2112/SI86-006.1

93. Lebel, P., Whangchai, N., Chitmanat, C., & Lebel, L. (2015). Risk of impacts from extreme weather and climate in river-based tilapia cage culture in Northern Thailand. International Journal of Global Warming, 8(4), 534–554. https://doi.org/10.1504/IJGW.2015.073054

94. Little, D., Bhujel, R., & Pham, T. (2003). Advanced nursing of mixed-sex and mono-sex tilapia (Oreochromis niloticus) fry, and its impact on subsequent growth in fertilized ponds. Aquaculture, 221, 265–276. https://doi.org/10.1016/S0044-8486(03)00008-5
95. Macintosh, D. J., Varghese, T. J., & Rao Satyanarayana, G. P. (1985). Hormonal sex reversal of wild-spawned tilapia in India. Journal of Fish Biology, 26(2), 87–94. https://doi.org/10.1111/j.1095-8649.1985.tb04245.x
96. Makori, A. J., Abuom, P. O., Kapiyo, R., Anyona, D. N., & Dida, G. O. (2017). Effects of water physico-chemical parameters on tilapia (Oreochromis niloticus) growth in earthen ponds in Teso North Sub-County, Busia County. Fisheries and Aquatic Sciences, 20(1), 30. https://doi.org/10.1186/s41240-017-0075-7
97. Marques, F. B., Watterson, A., da Rocha, A. F., & Cavalli, L. S. (2020). Overview of Brazilian aquaculture production. Aquaculture Research, 51(12), 4838–4845. https://doi.org/10.1111/are.14828
98. Martínez-Porchas, M., Martínez-Córdova, L. R., Porchas-Cornejo, M. A., & López-Elías, J. A. (2010). Shrimp polyculture: A potentially profitable, sustainable, but uncommon aquacultural practice. Reviews in Aquaculture, 2(2), 73–85. https://doi.org/10.1111/j.1753-5131.2010.01023.x
99. Menasveta, P., Piyatiratitivorakul, S., Rungsupa, S., Moree, N., & Fast, A. W. (1993). Gonadal maturation and reproductive performance of giant tiger prawn (Penaeus monodon Fabricius) from the Andaman Sea and pond-reared sources in Thailand. Aquaculture, 116(2), 191–198. https://doi.org/10.1016/0044-8486(93)90008-M
100. Modadugu, V. G., & Acosta, B. (2004). A review of global tilapia farming practices. Aquac. Asia, 9. https://doi.org/10.3109/9780203308905-5
101. Mohamed Abdelrahman, A. A. (2018). Effect of Feeding Frequency and Stocking Density on Tilapia Oreochromis Niloticus and Lettuce Lactuca Sativa Production in Aquaponics System under the UAE Condition and Business Enterprise Analysis. Biology Theses. Retrieved from https://scholarworks.uaeu.ac.ae/bio_theses/1
102. Mohammed, W., Sabry, M., Shehata, El-Nagar, G., Khater, A., & Radwan, M. (2014). Effect Of Female Weight On Reproductive Performance Of Nile Tilapia (Oreochromis niloticus).

103. Mondal, M. N., Shahin, J., Wahab, M. A., Asaduzzaman, M., & Yang, Y. (Eds.). (2010). Comparison between cage and pond production of Thai Climbing Perch (Anabas testudineus) and Tilapia (Oreochromis niloticus) under three management systems. Journal of the Bangladesh Agricultural University. https://doi.org/10.22004/ag.econ.208608

104. Munasinghe, M. N., Stephen, C., Abeynayake, P., & Abeygunawardena, I. S. (2010). Shrimp Farming Practices in the Puttallam District of Sri Lanka: Implications for Disease Control, Industry Sustainability, and Rural Development. Veterinary Medicine International, 2010, e679130. https://doi.org/10.4061/2010/679130

105. Nahid, S., Karim, E., Hasan, M. R., Shahabuddin, A. M., & Bhuyain, M. (2012). Different Farming Systems And Comparative Advantages Of Tilapia Over Other Commercial Aquaculture Species In Bangladesh. Bangladesh J. Prog. Sci. & Tech., 10, 093–096.

106. Na-Nakorn, U. (1992). Review Of Aquaculture Genetic Researches In Thailand. BIOTROPIA - The Southeast Asian Journal of Tropical Biology, (6). https://doi.org/10.11598/btb.1992.0.6.107

107. Nasr-Allah, A. M., Dickson, M. W., Al-Kenawy, D. A. R., Ahmed, M. F. M., & El-Naggar, G. O. (2014). Technical characteristics and economic performance of commercial tilapia hatcheries applying different management systems in Egypt. Aquaculture, 426–427, 222–230. https://doi.org/10.1016/j.aquaculture.2014.02.004

108. Navarro, R. D., Alves, S. M. B., & Ribeiro, T. N. L. R. (2021). Seminal features and reproductive index of Thai tilapia (Oreochromis sp.) on different reproductive periods. Acta Scientiarum. Biological Sciences, 43, e50828–e50828. https://doi.org/10.4025/actascibiolsci.v43i1.50828

109. Naylor, R. L., Hardy, R. W., Buschmann, A. H., Bush, S. R., Cao, L., Klinger, D. H., … Troell, M. (2021). A 20-year retrospective review of global aquaculture. Nature, 591(7851), 551–563. https://doi.org/10.1038/s41586-021-03308-6

110. Nguyen, N. H. (2016). Genetic improvement for important farmed aquaculture species with a reference to carp, tilapia and prawns in Asia: Achievements, lessons and challenges. Fish and Fisheries, 17(2), 483–506. https://doi.org/10.1111/faf.12122

111. Nobrega, R. O., Banze, J. F., Batista, R. O., & Fracalossi, D. M. (2020). Improving winter production of Nile tilapia: What can be

done? Aquaculture Reports, 18, 100453. https://doi.org/10.1016/j.aqrep.2020.100453

112. Ofori, J. K., Abban, E. K., Karikari, A. Y., & Brummett, R. E. (2010). Production Parameters and Economics of Small-Scale Tilapia Cage Aquaculture in the Volta Lake, Ghana. Journal of Applied Aquaculture, 22(4), 337–351. https://doi.org/10.1080/10454438.2010.527591
113. Paller, V. G. V., Corpuz, M. N. C., & Bandal, M. J. Z. (2017). Freshwater Fish Assemblages and Water Quality Parameters in Seven Lakes of San Pablo, Laguna, Philippines. Asian Journal of Biodiversity, 8(1). https://doi.org/10.7828/ajob.v8i1.995
114. Panase, P., Saenphet, S., & Saenphet, K. (2018). Biochemical and physiological responses of Nile tilapia Oreochromis niloticus Lin subjected to cold shock of water temperature. Aquaculture Reports, 11, 17–23. https://doi.org/10.1016/j.aqrep.2018.05.005
115. Paprocki, K., & Cons, J. (2014). Life in a shrimp zone: Aqua- and other cultures of Bangladesh's coastal landscape. The Journal of Peasant Studies, 41(6), 1109–1130. https://doi.org/10.1080/03066150.2014.937709
116. Pet, J. S., Gevers, G. J. M., Van Densen, W. L. T., & Vijverberg, J. (1996). Management options for a more complete utilization of the biological fish production in Sri Lankan reservoirs. Ecology of Freshwater Fish, 5(1), 1–14. https://doi.org/10.1111/j.1600-0633.1996.tb00032.x
117. Pet, J. S., & Piet, G. J. (1993). The consequences of habitat occupation and habitat overlap of the introduced tilapia Oreochromis mossambicus and indigenous fish species for fishery management in a Sri Lankan reservoir. Journal of Fish Biology, 43(sA), 193–208. https://doi.org/10.1111/j.1095-8649.1993.tb01187.x
118. Pethiyagoda, R. (1994). Threats to the indigenous freshwater fishes of Sri Lanka and remarks on their conservation. In A. Sasekumar, N. Marshall, & D. J. Macintosh (Eds.), Ecology and Conservation of Southeast Asian Marine and Freshwater Environments including Wetlands (pp. 189–201). Dordrecht: Springer Netherlands. https://doi.org/10.1007/978-94-011-0958-1_19
119. Ponzoni, R. W., Hamzah, A., Tan, S., & Kamaruzzaman, N. (2005). Genetic parameters and response to selection for live weight in the GIFT strain of Nile Tilapia (Oreochromis niloticus). Aquaculture, 247(1), 203–210. https://doi.org/10.1016/j.aquaculture.2005.02.020

120. Prabu, E., Rajagopalsamy, C. B. T., Ahilan, B., Jeevagan, I. J. M. A., & Renuhadevi, M. (2019). Tilapia – An Excellent Candidate Species for World Aquaculture: A Review. Annual Research & Review in Biology, 1–14. https://doi.org/10.9734/arrb/2019/v31i330052

121. Prabu, Elangovan, Felix, N., & Uma, A. (2021). Optimizing amino acid balance in fish meal-free diets for GIFT strain of Nile tilapia (Oreochromis niloticus) by deletion method. Aquaculture Nutrition, 27(4), 1031–1041. https://doi.org/10.1111/anu.13244

122. Pradhan, P. K., Paria, A., Yadav, M. K., Verma, D. K., Gupta, S., Swaminathan, T. R., … Lal, K. K. (2020). Susceptibility of Indian major carp Labeo rohita to tilapia lake virus. Aquaculture, 515, 734567. https://doi.org/10.1016/j.aquaculture.2019.734567

123. Ragasa, C., Agyakwah, S. K., Asmah, R., Mensah, E. T.-D., Amewu, S., & Oyih, M. (2022). Accelerating pond aquaculture development and resilience beyond COVID: Ensuring food and jobs in Ghana. Aquaculture, 547, 737476. https://doi.org/10.1016/j.aquaculture.2021.737476

124. Rahman, M. L., Shahjahan, M., & Ahmed, N. (2021). Tilapia Farming in Bangladesh: Adaptation to Climate Change. Sustainability, 13(14), 7657. https://doi.org/10.3390/su13147657

125. Raj, N. S., Swaminathan, T. R., Dharmaratnam, A., Raja, S. A., Ramraj, D., & Lal, K. K. (2019). Aeromonas veronii caused bilateral exophthalmia and mass mortality in cultured Nile tilapia, Oreochromis niloticus (L.) in India. Aquaculture, 512, 734278. https://doi.org/10.1016/j.aquaculture.2019.734278

126. Rand, J., & Tarp, F. (2009). Impact of an aquaculture extension project in Bangladesh. Journal of Development Effectiveness, 1(2), 130–146. https://doi.org/10.1080/19439340902918110

127. Rao, G. S., Imelda, J., Philipose, K. K., & Mojjada, S. K. (2013). Cage Aquaculture in India. Kochi: Central Marine Fisheries Research Institute. Retrieved from http://eprints.cmfri.org.in/9218/

128. Rao, M., Kumar, S. H., Kumar, S., Bedekar, M. K., Tripathi, G., & Kooloth Valappil, R. (2021). Microbiological investigation of Tilapia lake virus–associated mortalities in cage-farmed Oreochromis niloticus in India. Aquaculture International, 29(2), 511–526. https://doi.org/10.1007/s10499-020-00635-9

129. Rathore, S. S., Murthy, H. S., Girisha, S. K., Nithin, M. S., Nasren, S., Mamun, M. A. A., … Pai, M. (2021). Supplementation of nano□ selenium in fish diet: Impact on selenium assimilation and immune-regulated selenoproteome expression in monosex Nile tilapia (Oreochromis niloticus). Comparative Biochemistry and Physiology Part C: Toxicology & Pharmacology, 240, 108907. https://doi.org/10.1016/j.cbpc.2020.108907

130. Rezk, M. A., Ponzoni, R. W., Khaw, H. L., Kamel, E., Dawood, T., & John, G. (2009). Selective breeding for increased body weight in a synthetic breed of Egyptian Nile tilapia, Oreochromis niloticus: Response to selection and genetic parameters. Aquaculture, 293(3), 187–194. https://doi.org/10.1016/j.aquaculture.2009.03.019

131. Ritchie, H., & Roser, M. (2021). Biodiversity. Our World in Data. Retrieved from https://ourworldindata.org/fish-and-overfishing

132. Ruangpan, L., Kitao, T., & Yoshida, T. (1986). Protective efficacy of Aeromonas hydrophila vaccines in Nile tilapia. Veterinary Immunology and Immunopathology, 12(1), 345–350. https://doi.org/10.1016/0165-2427(86)90139-X

133. Saiful Islam, A. H. Md., Barman, B. K., & Murshed-e-Jahan, K. (2015). Adoption and impact of integrated rice–fish farming system in Bangladesh. Aquaculture, 447, 76–85. https://doi.org/10.1016/j.aquaculture.2015.01.006

134. Saiti, F., Jamu, D. M., Chisala, B., & Kambewa, P. (2007). Simulation of optimal harvesting strategies for small-scale mixed-sex tilapia (Oreochromis shiranus Boulenger 1896) ponds using a bio-economic model. Aquaculture Research, 38(4), 340–350. https://doi.org/10.1111/j.1365-2109.2007.01671.x

135. Shamsuzzaman, Md. M., Hoque Mozumder, M. M., Mitu, S. J., Ahamad, A. F., & Bhyuian, Md. S. (2020). The economic contribution of fish and fish trade in Bangladesh. Aquaculture and Fisheries, 5(4), 174–181. https://doi.org/10.1016/j.aaf.2020.01.001

136. Soto, E., Wang, R., Wiles, J., Baumgartner, W., Green, C., Plumb, J., & Hawke, J. (2015). Characterization of Isolates of Streptococcus agalactiae from Diseased Farmed and Wild Marine Fish from the U.S. Gulf Coast, Latin America, and Thailand. Journal of Aquatic Animal Health, 27(2), 123–134. https://doi.org/10.1080/08997659.2015.1032439

137. Srinivas, K. S. (2022). Export Oriented Aquaculture for Seafood Exports from India. In Impact of Climate Change on Hydrological Cycle, Ecosystem, Fisheries and Food Security. CRC Press.

138. Subrahmanyan, R. (1959). Studies on the phytoplankton of the west coast of India. Proceedings / Indian Academy of Sciences, 50(4), 189–252. https://doi.org/10.1007/BF03051925

139. Sumithra, T. G., Reshma, K. J., Anusree, V. N., Sayooj, P., Sharma, S. R. K., Suja, G., … Sanil, N. K. (2019). Pathological investigations of Vibrio vulnificus infection in Genetically Improved Farmed Tilapia (Oreochromis niloticus L.) cultured at a floating cage farm of India. Aquaculture, 511, 734217. https://doi.org/10.1016/j.aquaculture.2019.734217

140. Sumuduni, B. G. D., Munasinghe, D. H. N., Chandrarathna, W. P. R., & Amarasinghe, N. J. D. S. (2015). Seasonal variation and site specificity of external parasites in gold fish (Carassius auratus L.) and koi carp (Cyprinus carpio L.) in Rambadagalle Ornamental Fish Breeding Center, Sri Lanka. Ruhuna Journal of Science, 6(1). Retrieved from https://rjs.ruh.ac.lk/index.php/rjs/article/view/79

141. Suresh, A. V., & Lin, C. K. (1992). Tilapia culture in saline waters: A review. Aquaculture, 106(3), 201–226. https://doi.org/10.1016/0044-8486(92)90253-H

142. Tain, F. H., & Diana, J. S. (2007). Impacts of Extension Practice: Lessons From Small Farm-Based Aquaculture of Nile Tilapia in Northeastern Thailand. Society & Natural Resources, 20(7), 583–595. https://doi.org/10.1080/08941920601171824

143. Tandavanitj, S., Henocque, Y., Silpachai, D., Pimoljinda, J., & Nabhitabhata, J. (n.d.). Fisheries Co-management: An Experiential Account from Phang Nga Bay, Thailand.

144. Tayamen, M. M., Reyes, R. A., Danting, M. J., Mendoza, A. M., Marquez, E. B., Salguet, A. C., … Vera-Cruz, E. M. (2002). Tilapia broodstock development for saline waters in the Philippines. NAGA. Retrieved from https://digitalarchive.worldfishcenter.org/handle/20.500.12348/2285

145. Thilakaratne, I. D. S. I. P., Rajapaksha, G., Hewakopara, A., Rajapakse, R. P. V. J., & Faizal, A. C. M. (2003). Parasitic infections in freshwater ornamental fish in Sri Lanka. Diseases of Aquatic Organisms, 54(2), 157–162. https://doi.org/10.3354/dao054157

146. Thunjai, T., Boyd, C. E., & Boonyaratpalin, M. (2004). Bottom soil quality in Tilapia ponds of different age in Thailand. Aquaculture Research, 35(7), 698–705. https://doi.org/10.1111/j.1365-2109.2004.01072.x
147. Thushari, G. G. N., & Senevirathna, J. D. M. (2020). Plastic pollution in the marine environment. Heliyon, 6(8), e04709. https://doi.org/10.1016/j.heliyon.2020.e04709
148. Tidwell, J. H., & Bright, L. A. (2019). Freshwater Aquaculture. In B. Fath (Ed.), Encyclopedia of Ecology (Second Edition) (pp. 91–96). Oxford: Elsevier. https://doi.org/10.1016/B978-0-12-409548-9.10618-9
149. Toufique, K. A., & Belton, B. (2014). Is Aquaculture Pro-Poor? Empirical Evidence of Impacts on Fish Consumption in Bangladesh. World Development, 64, 609–620. https://doi.org/10.1016/j.worlddev.2014.06.035
150. Tran, N., Shikuku, K. M., Rossignoli, C. M., Barman, B. K., Cheong, K. C., Ali, M. S., & Benzie, J. A. H. (2021). Growth, yield and profitability of genetically improved farmed tilapia (GIFT) and non-GIFT strains in Bangladesh. Aquaculture, 536, 736486. https://doi.org/10.1016/j.aquaculture.2021.736486
151. Transportation of Fish in Bags. (n.d.). Retrieved April 27, 2022, from https://thefishsite.com/articles/transportation-of-fish-in-bags
152. Trọng, T. Q., Mulder, H. A., van Arendonk, J. A. M., & Komen, H. (2013). Heritability and genotype by environment interaction estimates for harvest weight, growth rate, and shape of Nile tilapia (Oreochromis niloticus) grown in river cage and VAC in Vietnam. Aquaculture, 384–387, 119–127. https://doi.org/10.1016/j.aquaculture.2012.12.022
153. Tseng, M.-C., Yang, D.-H., & Yen, T.-B. (2017). Comparative Study on Hatching Rate, Survival Rate, and Feminization of Onychostoma barbatulum (Pellegrin, 1908) at Different Temperatures and Examining Sex Change by Gonad and Karyotype Analyses. Zoological Studies, 56, e16. https://doi.org/10.6620/ZS.2017.56-16
154. Turcios, A. E., & Papenbrock, J. (2014). Sustainable Treatment of Aquaculture Effluents—What Can We Learn from the Past for the Future? Sustainability, 6(2), 836–856. https://doi.org/10.3390/su6020836

155. Ukenye, E. A., Taiwo, I. A., & Anyanwu, P. E. (2019). Morphological and genetic variation in Tilapia guineensis in West African coastal waters: A mini review. Biotechnology Reports, 24, e00362. https://doi.org/10.1016/j.btre.2019.e00362

156. Valenti, W. C., Barros, H. P., Moraes-Valenti, P., Bueno, G. W., & Cavalli, R. O. (2021). Aquaculture in Brazil: Past, present and future. Aquaculture Reports, 19, 100611. https://doi.org/10.1016/j.aqrep.2021.100611

157. Vineetha Valsalan, K. C., Kripa, V., Padua, S., Narayanakumar, R., & Vysakhan, P. (2020). Socio-economic analysis of finfish culture in cages in coastal waters of Kerala, southwest coast of India. Journal of the Marine Biological Association of India, 62(1), 21–26. Retrieved from http://mbai.org.in/php/journaldload.php?id=2537&bkid=124

158. Wang, J.-Q., Li, D., Dong, S., Wang, K., & Tian, X. (1998). Experimental studies on polyculture in closed shrimp ponds: I. Intensive polyculture of Chinese shrimp (Penaeus chinensis) with tilapia hybrids. Aquaculture, 163(1), 11–27. https://doi.org/10.1016/S0044-8486(98)00165-3

159. Wang, M., & Lu, M. (2016). Tilapia polyculture: A global review. Aquaculture Research, 47(8), 2363–2374. https://doi.org/10.1111/are.12708

160. Watanabe, W. O., Losordo, T. M., Fitzsimmons, K., & Hanley, F. (2002). Tilapia Production Systems in the Americas: Technological Advances, Trends, and Challenges. Reviews in Fisheries Science, 10(3–4), 465–498. https://doi.org/10.1080/20026491051758

161. Wati, L. A., Chang, W.-I., & Mustadjab, M. M. (2013). Competitiveness of Indonesian Shrimp Compare with Thailand Shrimp in Export Market. Wacana Journal of Social and Humanity Studies, 16(1), 24–31. Retrieved from https://wacana.ub.ac.id/index.php/wacana/article/view/270

162. Wijenayake, W. M. H. K., Jayasinghe, U. a. D., Amarasinghe, U. S., Athula, J. A., Pushpalatha, K. B. C., & De Silva, S. S. (2005). Culture-based fisheries in non-perennial reservoirs in Sri Lanka: Production and relative performance of stocked species. Fisheries Management and Ecology, 12(4), 249–258. https://doi.org/10.1111/j.1365-2400.2005.00447.x

163. World population projected to reach 9.8 billion in 2050, and 11.2 billion in 2100 | UN DESA | United Nations Department of Economic and Social Affairs. (n.d.). Retrieved April 27, 2022, from https://

www.un.org/development/desa/en/news/population/world-population-prospects-2017.html

164. Xiao, W., Chen, B., Wang, J., Zou, Z., Wang, C., Li, D., … Yang, H. (2022). Integration of mRNA and miRNA Profiling Reveals Heterosis in Oreochromis niloticus × O. aureus Hybrid Tilapia. Animals, 12(5), 640. https://doi.org/10.3390/ani12050640

165. Yamamoto, K., Igawa, K., Sugimoto, K., Yoshizawa, Y., Yanagiguchi, K., Ikeda, T., … Hayashi, Y. (2014). Biological Safety of Fish (Tilapia) Collagen. BioMed Research International, 2014, e630757. https://doi.org/10.1155/2014/630757

166. Yamkasem, J., Tattiyapong, P., Gorgoglione, B., & Surachetpong, W. (2021). Uncovering the first occurrence of Tilapia parvovirus in Thailand in tilapia during co-infection with Tilapia tilapinevirus. Transboundary and Emerging Diseases, 68(6), 3136–3144. https://doi.org/10.1111/tbed.14143

167. Yáñez, J. M., Joshi, R., & Yoshida, G. M. (2020). Genomics to accelerate genetic improvement in tilapia. Animal Genetics, 51(5), 658–674. https://doi.org/10.1111/age.12989

168. Yap, W. G., Villaluz, A. C., Soriano, M. G. G., & Santos, M. N. (n.d.). Milkfish Production and Processing Technologies in the Philippines. 97.

169. Yi, Y., Lin, C. K., & Diana, J. S. (1996). Influence of Nile tilapia (Oreochromis niloticus) stocking density in cages on their growth and yield in cages and in ponds containing the cages. Aquaculture, 146(3), 205–215. https://doi.org/10.1016/S0044-8486(96)01377-4

170. Zaman, T., Jewel, M. a. S., & Bhuiyan, A. S. (2006). Present status of pond fishery resources and livelihood of the fish farmers at Mohanpur Upazila in Rajshahi District. University Journal of Zoology, Rajshahi University, 25, 31–35. https://doi.org/10.3329/ujzru.v25i0.323

171. Zhang, Z. (2021). Research Advances on Tilapia Streptococcosis. Pathogens, 10(5), 558. https://doi.org/10.3390/pathogens10050558

INDEX

E

F

G

H

I

L

M

N